Lecture Notes in Business Information Processing 237

More information about this series at http://www.springer.com/series/7911

Paolo Ceravolo · Barbara Russo
Rafael Accorsi (Eds.)

Data-Driven Process Discovery and Analysis

4th International Symposium, SIMPDA 2014
Milan, Italy, November 19–21, 2014
Revised Selected Papers

 Springer

Editors
Paolo Ceravolo
Università degli Studi di Milano
Crema
Italy

Barbara Russo
Free University of Bozen/Bolzano
Bolzano
Italy

Rafael Accorsi
Universität Freiburg
Freiburg
Germany

ISSN 1865-1348 ISSN 1865-1356 (electronic)
Lecture Notes in Business Information Processing
ISBN 978-3-319-27242-9 ISBN 978-3-319-27243-6 (eBook)
DOI 10.1007/978-3-319-27243-6

Library of Congress Control Number: 2015955854

Springer Cham Heidelberg New York Dordrecht London

Springer International Publishing AG Switzerland is part of Springer Science+Business Media
(www.springer.com)

Preface

The rapid growth of organizational and business process managed via information systems made available a big variety of data that as a consequence created a high demand for making available data analysis techniques more effective and valuable. The fourth edition of the International Symposium on Data-Driven Process Discovery and Analysis (SIMPDA 2014) was conceived to offer a forum where researchers from different communities and the industry can share their insights in this hot new field. The symposium featured a number of advanced keynotes illustrating new approaches as well as presentations on recent research. The goal is to foster exchanges between academic researchers, industry, and a wider audience interested in process discovery and analysis. The event is organized by the IFIP WG 2.6. This year, the symposium was held in Milan, the city of Expo 2015.

The submissions cover theoretical issues related to process representation, discovery, and analysis or provide practical and operational experiences in process discovery and analysis. To improve the quality of the contributions, the symposium fostered discussion during the presentations, giving authors the opportunity to improve their work extending the results presented. For this reason, authors of accepted papers and keynote speakers were invited to submit extended articles to this proceedings volume in the LNBIP series. There were 21 submissions and five papers were accepted for publication.

During this edition the presentations and the discussions frequently focused on the implementation of process-mining algorithms in contexts where the analytical process is fed by data streams. The current selection of papers underlines the most relevant challenges that were identified and proposes novel solutions and approaches facing these challenges.

In the first paper "Discovery of Frequent Episodes in Event Logs," Maikel Leemans and Wil M.P. van der Aalst present an approach to detect frequently occurring episodes, i.e., a partially ordered collection of events, in an event log. Moreover, this work uses comparison with existing discovery algorithms to demonstrate that episode mining benefits from exploiting parameters that are encoding the process behavior.

The second paper by Bart Hompes et al., "Finding Suitable Activity Clusters for Decomposed Process Discovery," focused on decomposition as a strategy for the parallelization of process-mining algorithms. Analysis shows that although the decomposition step takes a relatively small amount of time, it is of key importance in finding a high-quality process model and for the computation time required to discover the individual parts. Moreover, the authors propose three metrics that can be used to assess the quality of a decomposition, before using it to discover a model or check conformance.

The third paper by Mahdi Alizadeh, Massimiliano de Leoni, and Nicola Zannone, "History-Based Construction of Alignments for Conformance Checking: Formalization and Implementation?", proposes an approach to automatically define the cost function for alignment-based conformance-checking techniques. Based on the information extracted from the past process executions, the cost function is derived relying on objective factors and thus enabling the construction of probable alignments, i.e., alignments that provide probable explanations of nonconformity.

The fourth paper, by David Redlich et al., "Dynamic Constructs Competition Miner – Occurrence vs. Time-Based Ageing," extends a divide-and-conquer algorithm for discovering block-structured processes from event logs possibly consisting of exceptional behavior. In particular, this paper proposes a set of modifications to enable dynamic business process discovery at run-time from a stream of events.

The fifth paper by Marco Anisetti et al., "Trustworthy Cloud Certification: A Model-Based Approach," discusses the problem of tracking and assessing the behavior of cloud services/processes. One of the main limitations of existing approaches is the uncertainty introduced by the cloud on the validity and correctness of existing certificates. The authors present a trustworthy cloud certification approach by continuously verifying the correctness of the service model at the basis of certification activities against real and synthetic service execution traces.

We gratefully acknowledge the strong research community that gathered around the research problems related to process data analysis and the high quality of their research work, which is hopefully reflected in the papers of this volume. We would also like to express our deep appreciation for the reviewers hard work and dedication. Above all, thanks are due the authors for submitting the best results of their work to the Symposium on Data-Driven Process Discovery and Analysis.

We are very grateful to the Università degli Studi di Milano and to IFIP for their financial support, and to the University of Freiburg and the Free University of Bozen/Bolzano.

October 2015

Paolo Ceravolo
Rafael Accorsi
Barbara Russo

Organization

Conference Co-chairs

Paolo Ceravolo   Università degli Studi di Milano, Italy
Rafael Accorsi   Universität Freiburg, Germany
Barbara Russo   Free University of Bozen/Bolzano, Italy

Advisory Board

Karl Aberer   EPFL, Switzerland
Ernesto Damiani   Università degli Studi di Milano, Italy
Tharam Dillon   La Trobe University, Australia
Dragan Gasevic   Athabasca University, Canada
Marcello Leida   EBTIC (Etisalat BT Innovation Centre), UAE
Erich Neuhold   University of Vienna, Austria
Maurice van Keulen   University of Twente, The Netherlands

PhD Award Committee

Paolo Ceravolo   Università degli Studi di Milano, Italy
Marcello Leida   EBTIC (Etisalat BT Innovation Centre), UAE
Gregorio Piccoli   Zucchetti spa, Italy

Web Chair and Publicity Chair

Fulvio Frati   Università degli Studi di Milano, Italy

Program Committee

Irene Vanderfeesten   Eindhoven University of Technology, The Netherlands
Maurice Van Keulen   University of Twente, The Netherlands
Manfred Reichert   University of Ulm, Germany
Schahram Dustdar   Vienna University of Technology, Austria
Mohamed Mosbah   University of Bordeaux, France
Meiko Jensen   Ruhr-University Bochum, Germany
Helen Balinsky   Hewlett-Packard Laboratories, UK
Valentina Emilia Balas   University of Arad, Romania
Karima Boudaoud   Ecole Polytechnique de Nice Sophia Antipolis, France
George Spanoudakis   City University London, UK
Richard Chbeir   University of Bourgogne, France
Gregorio Martinez Perez   University of Murcia, Spain

Contents

Discovery of Frequent Episodes in Event Logs

Maikel Leemans$^{(\boxtimes)}$ and Wil M.P. van der Aalst

Eindhoven University of Technology, P.O. Box 513,
5600 MB Eindhoven, The Netherlands
{m.leemans,w.m.p.v.d.aalst}@tue.nl

Abstract. Lion's share of process mining research focuses on the discovery of end-to-end process models describing the characteristic behavior of observed cases. The notion of a process instance (i.e., the case) plays an important role in process mining. Pattern mining techniques (such as traditional episode mining, i.e., mining collections of partially ordered events) do not consider process instances. In this paper, we present a new technique (and corresponding implementation) that discovers frequently occurring *episodes* in event logs, thereby exploiting the fact that events are associated with cases. Hence, the work can be positioned in-between process mining and pattern mining. Episode Discovery has its applications in, amongst others, discovering local patterns in complex processes and conformance checking based on partial orders. We also discover episode rules to predict behavior and discover correlated behaviors in processes, and apply our technique to other perspectives present in event logs. We have developed a ProM plug-in that exploits efficient algorithms for the discovery of frequent episodes and episode rules. Experimental results based on real-life event logs demonstrate the feasibility and usefulness of the approach.

Keywords: Episode discovery · Partial order discovery · Process discovery

1 Introduction

Process mining provides a powerful way to analyze operational processes based on event data. Unlike classical purely model-based approaches (e.g., simulation and verification), process mining is driven by "raw" observed behavior instead of assumptions or aggregate data. Unlike classical data-driven approaches, process mining is truly process-oriented and relates events to high-level end-to-end process models [1].

In this paper, we use ideas *from episode mining* [2] *and apply these to the discovery of partially ordered sets of activities in event logs*. Event logs serve as the starting point for process mining. An event log can be viewed as a multiset of *traces* [1]. Each trace describes the life-cycle of a particular *case* (i.e., a *process instance*) in terms of the *activities* executed. Often event logs store additional information about events, e.g., the *resource* (i.e., the person or device) executing or initiating the activity, the *timestamp* of the event, or *data elements* (e.g., cost or involved products) recorded with the event.

© IFIP International Federation for Information Processing 2015
P. Ceravolo et al. (Eds.): SIMPDA 2014, LNBIP 237, pp. 1–31, 2015.
DOI: 10.1007/978-3-319-27243-6_1

Each trace in the event log describes the life-cycle of a case from start to completion. Hence, process discovery techniques aim to transform these event logs into *end-to-end process models*. Often the overall end-to-end process model is rather complicated because of the variability of real life processes. This results in "Spaghetti-like" diagrams. Therefore, it is interesting to also search for more local patterns in the event log – using episode discovery – while still exploiting the notion of process instances. Another useful application of episode discovery is discovering patterns while using other perspectives also present the event log. Lastly, we can use episode discovery as a starting point for conformance checking based on partial orders [3].

Since the seminal papers related to the Apriori algorithm [4–6], many pattern mining techniques have been proposed. These techniques do not consider the ordering of events [4] or assume an unbounded stream of events [5,6] without considering process instances. Mannila et al. [2] proposed an extension of sequence mining [5,6] allowing for partially ordered events. An episode is a partially ordered set of activities and it is frequent if it is "embedded" in many sliding time windows. Unlike in [2], our episode discovery technique does not use an arbitrary sized sliding window. Instead, we exploit the notion of process instances. Although the idea is fairly straightforward, as far as we know, this notion of frequent episodes was never applied to event logs.

Numerous applications of process mining to real-life event logs illustrate that *concurrency* is a key notion in process discovery [1,7,8]. One should avoid showing all observed *interleavings* in a process model. First of all, the model gets too complex (think of the classical "state-explosion problem"). Second, the resulting model will be overfitting (typically one sees only a fraction of the possible interleavings). This makes the idea of episode mining particularly attractive.

The remainder of this paper is organized as follows. Section 2 positions the work in existing literature. The novel notion of episodes and the corresponding rules are defined in Sect. 3. Section 4 describes the algorithms and corresponding implementation in the process mining framework *ProM*, available through the *Episode Miner* package [9]. The approach and implementation are evaluated in Sect. 5 using several publicly available event logs. Section 6 concludes the paper.

2   Related Work

The notion of frequent episode mining was first defined by Mannila et al. [2]. In their paper, they applied the notion of frequent episodes to (large) event sequences. The basic pruning technique employed in [2] is based on the frequency of episodes in an event sequence. Mannila et al. considered the mining of serial and parallel episodes separately, each discovered by a distinct algorithm. Laxman and Sastry improved on the episode discovery algorithm of Mannila by employing new frequency calculation and pruning techniques [10]. Experiments suggest that the improvement of Laxman and Sastry yields a 7 times speedup factor on both real and synthetic datasets.

Related to the discovery of episodes or partial orders is the discovery of end-to-end process models able to capture concurrency explicitly. The α algorithm [11]

Table 1. Feature comparison of discussed discovery algorithms

	Exploits process instances	Discovers end-to-end model	Focus on local behavior	Soundness guaranteed	Sequence	Choice	Concurrency	Silent (tau) transitions	Duplicate Activities
Agrawal, Sequence mining [4]	-	-	-	n.a.	+	-	-	-	-
Manilla, Episode mining [2]	-	-	+	n.a.	+	-	+	-	-
Leemans M., Episode Discovery	+	-	+	n.a.	+	-	+	-	+
Maggi, DECLARE Miner [23, 24, 25]	+	+/-	-	n.a.	+	+	+	-	+
Van der Aalst, α-algorithm [11]	+	+	-	-	+	+	+	-	-
Weijters, Heuristics mining [14]	+	+	-	-	+	+	+	-	-
De Medeiros, Genetic mining [15, 16]	+	+	-	-	+	+	+	+	+
Solé, State Regions [17, 18]	+	+	-	-	+	+	+	-	-
Bergenthum, Language Regions [19, 20]	+	+	-	-	+	+	+	-	-
Günther, Fuzzy mining [21]	+	+	-	n.a.	+	+/-	+/-	-	-
Leemans S.J.J., Inductive [22]	+	+	-	+	+	+	+	+	-

was the first process discovery algorithm adequately handling concurrency. Several variants of the α algorithm have been proposed [12,13]. Many other discovery techniques followed, e.g., *heuristic* mining [14] able to deal with noise and low-frequent behavior. The HeuristicsMiner is based on the notion of causal nets (C-nets). Moreover, completely different approaches have been proposed, e.g., the different types of *genetic* process mining [15,16], techniques based on *state-based regions* [17,18], and techniques based on *language-based regions* [19,20]. A frequency-based approach is used in the *fuzzy* mining technique, which produces a precedence-relation-based process map [21]. Frequencies are used to filter out infrequent paths and nodes. Another, more recent, approach is *inductive* process mining where the event log is split recursively [22]. The latter technique always produces a block-structured and sound process model. All the discovery techniques mentioned are able to uncover concurrency based on example behavior in the log. Additional feature comparisons are summarized in Table 1. Based on the above discussion we conclude that *Episode Discovery* is the only technique whose results focus on local behavior while exploiting process instances.

The discovery of Declarative Process Models, as presented in [23–25], aims to discover patterns to describe an overall process model. The underlying model is the DECLARE declarative language. This language uses LTL templates that can be used to express rules related to the ordering and presence of activities. This discovery technique requires the user to limit the constraint search-space by selecting rule templates to search for. That is, the user selects a subset of pattern types (e.g., succession, not-coexists, etc.) to search for. However, the underlying discovery technique is pattern-agnostic, and simply generates all pattern instantiations (using apriori-based optimization techniques), followed by LTL evaluations. The major downside of this approach is a relatively bad runtime performance, and we will also observe this in Sect. 5.4.

The discovery of patterns in the resource perspective has been partly tackled by techniques for organizational mining [26]. These techniques can be used to discover organizational models and social networks. A social network is a graph/network in which the vertices represent resources (i.e., a person or device), and the edges denote the relationship between resources. A typical example is the handover of work metric. This metric captures that, if there are two subsequent events in a trace, which are completed by resource a and b respectively, then it is likely that there is a handover of work from a to b. In essence, the discovery of handover of work network yields the "end-to-end" resource model, related to the discovery of episodes or partial orders on the resource perspective.

The episode mining technique presented in this paper is based on the discovery of frequent item sets. A well-known algorithm for mining frequent item sets and association rules is the Apriori algorithm by Agrawal and Srikant [4]. One of the pitfalls in association rule mining is the huge number of solutions. One way of dealing with this problem is the notion of representative association rules, as described by Kryszkiewicz [27]. This notion uses user specified constraints to reduce the number of 'similar' results. Both sequence mining [5,6] and episode mining [2] can be viewed as extensions of frequent item set mining.

3   Definitions: Event Logs, Episodes, and Episode Rules

This section defines basic notions such as event logs, episodes and rules. Note that our notion of episodes is different from the notion in [2] which does not consider process instances.

3.1   Preliminaries

Multisets. Multisets are used to describe event logs where the same trace may appear multiple times.

We denote the set of all multisets over some set A as $\mathcal{B}(A)$. We define $B(a)$ for some multiset $B \in \mathcal{B}(A)$ as the number of times element $a \in A$ appears in multiset B. For example, given $A = \{x, y, z\}$, a possible multiset $B \in \mathcal{B}(A)$ is $B = [x, x, y]$. For this example, we have $B(x) = 2$, $B(y) = 1$ and $B(z) = 0$. The size $|B|$ of a multiset $B \in \mathcal{B}(A)$ is the sum of appearances of all elements in the multiset, i.e.: $|B| = \Sigma_{a \in A} B(a)$.

Note that the ordering of elements in a multiset is irrelevant.

Sequences. Sequences are used to represent traces in an event log.

Given a set X, a sequence over X of length n is denoted as $\sigma = \langle a_1, a_2, \ldots, a_n \rangle \in X^*$. We denote the empty sequence as $\langle \rangle$.

Note that the ordering of elements in a sequence is relevant.

Functions. Given sets X and Y, we write $f : X \mapsto Y$ for the function with domain $\mathbf{dom}\, f \subseteq X$ and range $\mathbf{ran}\, f = \{\, f(x) \mid x \in X \,\} \subseteq Y$. In this context, the $\mapsto$ symbol is used to denote a specific function.

As an example, the function $f : \mathbb{N} \mapsto \mathbb{N}$ can be defined as $f = \{ x \mapsto x + 1 \mid x \in \mathbb{N} \}$. For this f we have, amongst others, $f(0) = 1$ and $f(1) = 2$ (i.e., this f defines a succession relation on $\mathbb{N}$).

3.2   Event Logs

Activities and Traces. Let $\mathcal{A} \subseteq \mathcal{U_A}$ be the alphabet of activities occurring in the event log. A trace is a sequence $\sigma = \langle a_1, a_2, \ldots, a_n \rangle \in \mathcal{A}^*$ of activities $a_i \in \mathcal{A}$ occurring at time index i relative to the other activities in σ.

Event Log. An event log $L \in \mathcal{B}(\mathcal{A}^*)$ is a multiset of traces. Note that the same trace may appear multiple times in an event log. Each trace corresponds to an execution of a process, i.e., a *case* or *process instance*. In this simple definition of an event log, an event refers to just an *activity*. Often event logs store additional information about events, such as the *resource* (i.e., the person or device) executing or initiating the activity, and the *timestamp* of the event.

Note that, in this paper, we assumed simple event logs using the default activity classifier, yielding partial orders on activities. It should be noted that the technique discussed in this paper is classifier-agnostic. As a result, using alternative classifiers, partial orders on other perspectives can be obtained. An example is the flow of work between persons by discovering partial orders using a resource classifier on the event log.

3.3   Episodes

Episode. An episode is a partially ordered collection of events. A partial order is a binary relation which is reflexive, antisymmetric and transitive. Episodes are depicted using the transitive reduction of directed acyclic graphs, where the nodes represent events, and the edges imply the partial order on events. Note that the presence of an edge implies serial behavior. Figure 1 shows the transitive reduction of an example episode.

Formally, an episode $\alpha = (V, \leq, g)$ is a triple, where V is a set of events (nodes), $\leq$ is a partial order on V, and $g : V \mapsto \mathcal{A}$ is a left-total function from events to activities, thereby labeling the nodes/events [2]. For two vertices $u, v \in V$ we have $u < v$ iff $u \leq v$ and $u \neq v$.

Note that if $|V| \leq 1$, then we got a singleton or empty episode. For the rest of this paper, we ignore empty episodes. We call an episode *parallel* when there are two or more vertices, and no edges.

Subepisode and Equality. An episode $\beta = (V', \leq', g')$ is a subepisode of $\alpha = (V, \leq, g)$, denoted $\beta \preceq \alpha$, iff there is an injective mapping $f : V' \mapsto V$ such that:

$$(\forall v \in V' : g'(v) = g(f(v))) \qquad \text{All vertices in } \beta \text{ are also in } \alpha$$
$$\wedge\, (\forall v, w \in V' \wedge v \leq' w : f(v) \leq f(w)) \qquad \text{All edges in } \beta \text{ are also in } \alpha$$

An episode β equals episode α, denoted $\beta \equiv \alpha$ iff $\beta \preceq \alpha \wedge \alpha \preceq \beta$. An episode β is a strict subepisode of α, denoted $\beta \prec \alpha$, iff $\beta \preceq \alpha \wedge \beta \not\equiv \alpha$.

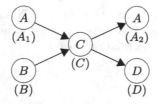

Fig. 1. Shown is the transitive reduction of the partial order for an example episode. The circles represent nodes (events), with the activity labeling imposed by g inside the circles, and an event ID beneath the nodes in parenthesis. In this example, events A_1 and B can happen in parallel (as can A_2 and D). However, event C can only happen after both an A_1 and a B have occurred, and A_2 and D can only happen after an C has occurred.

Episode Construction. Two episodes $\alpha = (V, \leq, g)$ and $\beta = (V', \leq', g')$ can be 'merged' to construct a new episode $\gamma = (V'', \leq'', g'')$. $\alpha \oplus \beta$ is a smallest γ (i.e., smallest sets V'' and $\leq''$) such that $\alpha \preceq \gamma$ and $\beta \preceq \gamma$.

The smallest sets criterion implies that every event $v \in V''$ and ordered pair $v, w \in V'' \wedge v \leq'' w$ must be represented in α and/or β (i.e., have a witness, see also the formulae below). Formally, an episode $\gamma = \alpha \oplus \beta$ iff there exists injective mappings $f : V \mapsto V''$ and $f' : V' \mapsto V''$ such that:

$$\gamma = (V'', \leq'', g'')$$
$$\leq'' = \{\, (f(v), f(w)) \mid (v, w) \in\ \leq \}$$
$$\cup \{\, (f'(v), f'(w)) \mid (v, w) \in\ \leq' \,\} \qquad\qquad \text{order witness}$$
$$g'' : (\forall v \in V : g(v) = g''(f(v))) \wedge (\forall v' \in V' : g'(v') = g''(f'(v'))) \quad \text{correct mapping}$$
$$V'' : \forall v'' \in V'' : (\exists v \in V : f(v) = v'') \vee (\exists v' \in V' : f'(v') = v'') \qquad \text{node witness}$$

Observe that "order witness" and "correct mapping" are based on $\alpha \preceq \gamma$ and $\beta \preceq \gamma$. Note that via "note witness" it is ensured that every vertex in V'' is mapped to a vertex in either V or V'. Every vertex in V and V' should be mapped to a vertex in V''. This is ensured via "correct mapping".

Occurrence. An episode $\alpha = (V, \leq, g)$ occurs in an event trace $\sigma = \langle a_1, a_2, \ldots, a_n \rangle$, denoted $\alpha \sqsubseteq \sigma$, iff there exists an injective mapping $h : V \mapsto \{1, .., n\}$ such that:

$$(\forall v \in V : g(v) = a_{h(v)} \in \sigma) \qquad\qquad \text{All vertices are mapped correctly}$$
$$\wedge\ (\forall v, w \in V \wedge v \leq w : h(v) \leq h(w)) \qquad \text{The partial order } \leq \text{ is respected}$$

In Fig. 2 an example of an "event to trace map" h for occurrence checking is given. Note that multiple mappings might exists. Intuitively, if we have a trace t and an episode with $u \leq v$, then the activity $g(u)$ must occur before activity $g(v)$ in t.

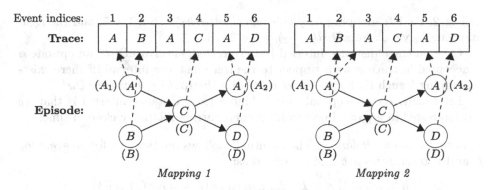

Fig. 2. Shown are two possible mappings h (the dotted arrows) for checking occurrence of the example episode in a trace. The shown graphs are the transitive reduction of the partial order of the example episode. Note that with the left mapping (*Mapping 1*) also an episode with the partial order $A_1 < B$ occurs in the given trace, in the right mapping (*Mapping 2*) the same holds for an episode with the partial order $B < A_1$.

Frequency. The frequency $freq(\alpha)$ of an episode α in an event log $L \in \mathcal{B}(\mathcal{A}^*)$ is defined as:

$$freq(\alpha) = \frac{|[\sigma \in L \mid \alpha \sqsubseteq \sigma]|}{|L|}$$

Given a frequency threshold $minFreq$, an episode α is frequent iff $freq(\alpha) \geq minFreq$. During the actual episode discovery, we use the contrapositive of the fact given in Lemma 1. That is, we use the observation that if not all subepisodes β are frequent, then the episode α is also not frequent.

Lemma 1 (Frequency and subepisodes). *If an episode α is frequent in an event log L, then all subepisodes β with $\beta \preceq \alpha$ are also frequent in L. Formally, we have for a given α:*

$$(\forall \beta \preceq \alpha : freq(\beta) \geq freq(\alpha))$$

3.4   Episode and Event Log Measurements

Activity Frequency. The activity frequency $ActFreq(a)$ of an activity $a \in \mathcal{A}$ in an event log $L \in \mathcal{B}(\mathcal{A}^*)$ is defined as:

$$ActFreq(a) = \frac{|[\sigma \in L \mid a \in \sigma]|}{|L|}$$

Given a frequency threshold $minActFreq$, an activity a is frequent iff $ActFreq(a) \geq minActFreq$.

Trace Distance. Given episode $\alpha = (V, \leq, g)$ occurring in an event trace $\sigma = \langle a_1, a_2, \ldots, a_n \rangle$, as indicated by an event to trace map $h : V \mapsto \{1, .., n\}$. Then the trace distance $traceDist(\alpha, h)$ is defined as:

$$traceDist(\alpha, h) = \max \{ h(v) \mid v \in V \} - \min \{ h(v) \mid v \in V \}$$

In Fig. 2, the left mapping h_1 yields $traceDist(\alpha, h_1) = 6 - 1 = 5$, and the right mapping h_2 yields $traceDist(\alpha, h_2) = 6 - 2 = 4$.

Given a trace distance interval $[minTraceDist, maxTraceDist]$, an episode α is accepted in trace σ with respect to the trace distance interval iff there exists a mapping h such that $minTraceDist \leq traceDist(\alpha, h) \leq maxTraceDist$.

Informally, the conceptual idea behind a trace distance interval is that we are interested in a partial order on events occurring relatively close in time.

Eventually-follows Relation. The eventually-follows relation $\gg_L$ for an event log L and two activities $a, b \in \mathcal{A}$ is defined as:

$$a \gg_L b = \left| \left\{ \sigma \in L \mid \exists_{0 \leq i < j < |\sigma|} : \sigma(i) = a \wedge \sigma(j) = b \right\} \right|$$

Informally, the eventually-follows valuation for $a \gg_L b$ equals the amount of traces in which a happens (at timestamp i), and is followed by b at a later moment (at timestamp j with $i < j$).

If we evaluate the eventually-follows relation for every $a, b \in \mathcal{A}$, we obtain the eventually-follows matrix. In Table 2 the eventually-follows matrix is given for an example event log.

Lemma 2 (Eventually-follows Relation and Episode Frequency). *The eventually-follows valuation $g(u) \gg_L g(v)$ for any two vertices $u, v \in V$ with $u \leq v$ is an upper bound for the frequency of the episode $\alpha = (V, \leq, g)$ in event log L. Formally:*

$$\left(\forall u, v \in V \wedge u \leq v : \frac{g(u) \gg_L g(v)}{|L|} \geq freq(\alpha) \right)$$

Consequently, if an episode $\alpha = (V, \leq, g)$ is frequent in an event log L, then for any two vertices $u, v \in V$ with $u \leq v$ also the eventually follows valuation for $g(u) \gg_L g(v)$ is frequent.

Based on Lemma 2, the eventually-follows relation can be used as a fast approximation of early occurrence checking. Concretely, by contraposition, we know that if there exists $u, v \in V$ with $u \leq v$ for which $\frac{g(u) \gg_L g(v)}{|L|} < minFreq$, then the episode α cannot be frequent. We use this fact as an optimization technique in the realization of our Episode Discovery technique.

Table 2. The eventually-follows matrix for the following example event log: $L = [\langle a, b, a, c, a, d \rangle, \langle a, b, a, d \rangle, \langle b, d \rangle]$. Each cell gives the valuation for *row* $\gg_L$ *column*, where *row* is the activity shown to the left, and *column* is the activity shown on the top of the table.

$\gg_L$	a	b	c	d
a	2	2	1	2
b	2	0	1	3
c	1	0	0	1
d	0	0	0	0

3.5 Episode Rules

Episode Rule. An episode rule is an association rule $\beta \Rightarrow \alpha$ with $\beta \prec \alpha$ stating that after seeing β, then likely the larger episode α will occur as well.

The confidence of the episode rule $\beta \Rightarrow \alpha$ is given by:

$$conf(\beta \Rightarrow \alpha) = \frac{freq(\alpha)}{freq(\beta)}$$

Given a confidence threshold *minConf*, an episode rule $\beta \Rightarrow \alpha$ is valid iff $conf(\beta \Rightarrow \alpha) \geq minConf$. During the actual episode rule discovery, we use Lemma 3.

Lemma 3 (Confidence and subepisodes). *If an episode rule $\beta \Rightarrow \alpha$ is valid in an event log L, then for all episodes β' with $\beta \prec \beta' \prec \alpha$ the event rule $\beta' \Rightarrow \alpha$ is also valid in L. Formally:*

$$(\forall \beta \prec \beta' \prec \alpha : conf(\beta \Rightarrow \alpha) \leq conf(\beta' \Rightarrow \alpha))$$

Episode Rule Magnitude. Let the graph size $size(\alpha)$ of an episode α be denoted as the sum of the nodes and edges in the transitive reduction of the episode. The magnitude of an episode rule is defined as:

$$mag(\beta \Rightarrow \alpha) = \frac{size(\beta)}{size(\alpha)}$$

Intuitively, the magnitude of an episode rule $\beta \Rightarrow \alpha$ represents how much episode α 'adds to' or 'magnifies' episode β. The magnitude of an episode rule allows smart filtering on generated rules. Typically, an extremely low (approaching zero) or high (approaching one) magnitude indicates a trivial episode rule.

4 Realization

The definitions and insights provided in the previous section have been used to implement an episode (rule) discovery plug-in in the process mining framework *ProM*, available through the *Episode Miner* package [9]. To be able to analyze real-life event logs, we need efficient algorithms. These are described next.

4.1 Notation in Realization

In the listed algorithms, we will reference to the elements of an episode $\alpha = (V, \leq, g)$ as $\alpha.V$, $\alpha.\leq$ and $\alpha.g$.

For the implementation, we rely on *ordered* sets, i.e., lists of unique elements. The order of a set is determined by the order in which elements are added to the sets, which is leveraged to make the algorithms efficient. We assume individual elements can be accessed via an index, with indexing starting at zero. We use the following operations and notations in the algorithms to come:

$$
\begin{aligned}
A &= \{x, y, z\} \text{ with } x < y < z & \text{Note: } n = |A| = 3 \\
A[0] &= x & \text{Access the first element} \\
A[n-1] &= z & \text{Access the last element} \\
(A \cup \{v\}) &= \{x, y, z, v\} \text{ with } x < y < z < v & \text{Adding new elements to a set} \\
(A \cup \{x\}) &= A & \text{Every element is unique} \\
(A \cup \{v\})[n] &= v & \text{Access the new last element} \\
A[0..n-2] &= \{x, y\} \text{ with } x < y & \text{Access a subset of a set}
\end{aligned}
$$

4.2 Frequent Episode Discovery

Discovering frequent episodes is done in two phases. The first phase discovers parallel episodes (i.e., nodes only); the second phase discovers partial orders (i.e., adding the edges). The main routine for discovering frequent episodes is given in Algorithm 1.

Algorithm 1. Episode Discovery

Input: An event log L, an activity alphabet $\mathcal{A}$, a frequency threshold $minFreq$.
Output: A set of frequent episodes Γ
Description: Two-phase episode discovery. Each phase alternates between recognizing frequent candidates in the event log (F_l), and generating new candidate episodes (C_l).
Proof of termination: Note that candidate episode generation with $F_l = \emptyset$ will yield $C_l = \emptyset$. Since each iteration the generated episodes become strictly larger (in terms of V and $\leq$), eventually the generated episodes cannot occur in any trace. Therefore, always eventually $F_l = \emptyset$, and thus we will always terminate.
EPISODEDISCOVERY(L, $\mathcal{A}$, $minFreq$)
(1) $\Gamma = \emptyset$
(2)   // Phase 1: discover parallel episodes
(3) $l = 1$ // Tracks the number of nodes
(4)   // Initialize: create a candidate episode for every activity in $\mathcal{A}$
(5) $C_l = \{ (V, \leq, g) \mid |V| = 1, \leq = \emptyset, g = \{v \mapsto a\}, v \in V, a \in \mathcal{A} \}$
(6)   // Step: recognize and construct larger episodes from smaller episodes
(7)   **while** $C_l \neq \emptyset$
(8) $F_l = $ RECOGNIZEFREQUENTEPISODES(L, C_l, $minFreq$)
(9) $\Gamma = \Gamma \cup F_l$
(10) $C_l = $ GENERATECANDIDATEPARALLEL(l, F_l)
(11) $l = l + 1$
(12)   // Phase 2: discover partial orders
(13) $l = 1$ // Tracks the number of edges
(14)   // Initialize: create candidate episodes based on results from Phase 1
(15) $C_l = \{ (\gamma.V, \leq, \gamma.g) \mid \gamma \in \Gamma, \leq = \{(v, w)\}, v, w \in \gamma.V, v \neq w \}$
(16)   // Step: recognize and construct larger episodes from smaller episodes
(17)   **while** $C_l \neq \emptyset$
(18) $F_l = $ RECOGNIZEFREQUENTEPISODES(L, C_l, $minFreq$)
(19) $\Gamma = \Gamma \cup F_l$
(20) $C_l = $ GENERATECANDIDATEORDER(l, F_l)
(21) $l = l + 1$
(22)   **return** Γ

4.3   Episode Candidate Generation

The generation of candidate episodes for each phase is an adaptation of the well-known Apriori algorithm over an event log. Given a set of frequent episodes F_l, we can construct a candidate episode γ by combining two partially overlapping episodes α and β from F_l. Note that this implements the episode construction operation $\gamma = \alpha \oplus \beta$.

For phase 1, we have F_l contains frequent episodes with l nodes and no edges. A candidate episode γ will have $l+1$ nodes, resulting from episodes α and β that overlap on the first $l-1$ nodes. This generation is implemented by Algorithm 2.

For phase 2, we have F_l contains frequent episodes with l edges. A candidate episode γ will have $l+1$ edges, resulting from episodes α and β that overlap on the

Algorithm 2. Candidate episode generation – Parallel

Input: A set of frequent episodes F_l with l nodes.
Output: A set of candidate episodes C_{l+1} with $l+1$ nodes.
Description: Generates candidate episodes γ by merging overlapping episodes α and β (i.e., $\gamma = \alpha \oplus \beta$). For parallel episodes, overlapping means: sharing $l-1$ nodes.
GENERATECANDIDATEPARALLEL(l, F_l)

```
(1)        C_{l+1} = ∅
(2)        for i = 0 to |F_l| − 1
(3)            for j = i to |F_l| − 1
(4)                α = F_l[i]
(5)                β = F_l[j]
(6)                // Check if α and β overlap (see also description, index
                   start at 0)
(7)                if ∀0 ≤ i ≤ l − 2 : α.g(α.V[i]) = β.g(β.V[i])
(8)                    // Create candidate γ = α ⊕ β
(9)                    γ = (V, ≤, g) where V = (α.V[0..l−1]∪β.V[l−1]), ≤ =
                       ∅, g = α.g ∪ β.g
(10)                   C_{l+1} = C_{l+1} ∪ {γ}
(11)               else
(12)                   break
(13)       return C_{l+1}
```

Algorithm 3. Candidate episode generation – Partial order

Input: A set of frequent episodes F_l with l edges.
Output: A set of candidate episodes C_{l+1} with $l+1$ edges.
Description: Generates candidate episodes γ by merging overlapping episodes α and β (i.e., $\gamma = \alpha \oplus \beta$). For partial order episodes, overlapping means: sharing all nodes and $l-1$ edges.
GENERATECANDIDATEORDER(l, F_l)

```
(1)        C_{l+1} = ∅
(2)        for i = 0 to |F_l| − 1
(3)            for j = i + 1 to |F_l| − 1
(4)                α = F_l[i]
(5)                β = F_l[j]
(6)                // Check if α and β overlap (see also description, index
                   start at 0)
(7)                sharingAllNodes = (α.V = β.V ∧ α.g = β.g)
(8)                overlappingEdges = (α.≤[0..l − 2] = β.≤[0..l − 2])
(9)                if sharingAllNodes ∧ overlappingEdges
(10)                   // Create candidate γ = α ⊕ β
(11)                   γ = (α.V, ≤, α.g) where ≤ = (α.E[0..l − 1] ∪ β.E[l − 1])
(12)                   C_{l+1} = C_{l+1} ∪ {γ}
(13)               else
(14)                   break
(15)       return C_{l+1}
```

first $l - 1$ edges and have the same set of nodes. This generation is implemented by Algorithm 3. Note that, formally, the partial order $\leq$ is the transitive closure of the set of edges being constructed.

4.4 Frequent Episode Recognition

In order to check if a candidate episode α is frequent, we check if $freq(\alpha) \geq minFreq$. The computation of $freq(\alpha)$ boils down to counting the number of traces σ with $\alpha \sqsubseteq \sigma$. Algorithm 4 recognizes all frequent episodes from a set of candidate episodes using the above described approach. Note that for both parallel and partial order episodes we can use the same recognition algorithm.

Recall that an event log is a multiset of traces. Based on this observation, we note that particular trace variants typically occur more than once in an event log. We use this fact to reduce the number of iterations in Algorithm 4, and consequently the number of occurrence checks performed (i.e., OCCURS() invocations). Instead of iterating over all the process instances on line 2 of the algorithm, we consider each trace variant σ only once. For the support count we use the $L(\sigma)$ multiset operation to get the correct number of process instances.

Algorithm 4. Recognize frequent episodes

Input: An event log L, a set of candidate episodes C_l, a frequency threshold $minFreq$.
Output: A set of frequent episodes F_l
Description: Recognizes frequent episodes, by filtering out candidate episodes that do not occur frequently in the log.
Note: If $F_l = \emptyset$, then $C_l = \emptyset$.
RECOGNIZEFREQUENTEPISODES($L, C_l, minFreq$)
(1) $support = [0, \ldots, 0]$ with $|support| = |C_l|$
(2)   foreach $\sigma \in L$
(3)   for $i = 0$ to $|C_l| - 1$
(4)   if OCCURS($C_l[i], \sigma$) then $support[i] = support[i] + L(\sigma)$
(5) $F_l = \emptyset$
(6)   for $i = 0$ to $|C_l| - 1$
(7)   if $\frac{support[i]}{|L|} \geq minFreq$ then $F_l = F_l \cup \{C_l[i]\}$
(8)   return F_l

Checking whether an episode α occurs in a trace $\sigma = \langle a_1, a_2, \ldots, a_n \rangle$ is done via checking the existence of the mapping $h : \alpha.V \mapsto \{1, .., n\}$. This results in checking the two propositions shown below. Algorithm 5 implements these checks.

– Checking whether each node $v \in \alpha.V$ has a unique witness in trace σ.
– Checking whether the (injective) mapping h respects the partial order indicated by $\alpha.\leq$.

Algorithm 5. Occurrence checking for an episode

Input: An episode α, a trace σ.
Output: True iff $\alpha \sqsubseteq \sigma$
Description: Implements occurrence checking based on finding an occurrence proof in the form of a mapping $h : \alpha.V \mapsto \{1, .., n\}$.
OCCURS($\alpha = (V, \leq, g), \sigma$)
(1)   // H indicates for each activity a all the indices i at which $a = a_i \in \sigma$
(2) $H = \{ a \mapsto \{ i \mid a = a_i \in \sigma \} \mid a \in \mathcal{A} \}$
(3) $h = \emptyset$
(4)   **return** CHECKMODEL(α, H, h)

Algorithm 6. This algorithm implements occurrence checking via recursive discovery of the injective mapping h as per the occurrence definition.

Input: An episode α, a class of mappings $H : \mathcal{A} \mapsto \mathcal{P}(\mathbb{N})$, and an intermediate mapping $h : \alpha.V \mapsto \{1, .., n\}$.
Output: True iff there is a mapping h, as per the occurrence definition, derivable from H
Description: Recursive implementation for finding h based on induction to the number of mapped vertices:
Base case (*if*-part): Every $v \in V$ is mapped ($v \in$ **dom** h).
Step case (*else*-part): (IH) n vertices are mapped, step by adding a mapping for a vertex $v \notin$ **dom** h.
CHECKMODEL($\alpha = (V, \leq, g), H, h$)
(1)   **if** $\forall v \in V : v \in$ **dom** h
(2)   // Every $v \in V$ is mapped, check the edge relation
(3)   **return** $(\forall (v, w) \in \leq : h(v) \leq h(w))$
(4)   **else**
(5)   // Choose a mapping for a vertex $v \notin$ **dom** h
(6)   **pick** $v \in V$ with $v \notin$ **dom** h
(7)   // Compute $\exists i \in H(g(v)) :$ CHECKMODEL(v mapped to i)
(8) $exists = False$
(9)   **toreach** $i \in H(g(v))$ **do** $exists \lor$ CHECKMODEL($\alpha, H\lfloor g(v) \mapsto H(g(v)) \setminus \{i\}\rfloor, h[v \mapsto i]$)
(10)   **return** $exists$

For the discovery of an injective mapping h for a specific episode α and trace σ we use the following recipe. First, we declare the class of models $H : \mathcal{A} \mapsto \mathcal{P}(\mathbb{N})$ such that for each activity $a \in \mathcal{A}$ we get the set of indices i at which $a = a_i \in \sigma$. Next, we try all possible models derivable from H. A model $h : \alpha.V \mapsto \{1, .., n\}$ is derived from H by choosing an index $i \in H(f(v))$ for each node $v \in \alpha.V$. With such a model h, we can perform the actual partial order check against $\alpha. \leq$.

4.5   Time Complexity Analysis

The theoretical time complexity of the provided algorithms is dominated by two aspects: (1) the Apriori-style iterations in Algorithm 1, and (2) the occurrence checking in Algorithm 6. For the worst case time complexity we will first investigate the occurrence checking, and then briefly display the total time complexity.

Analysis of Occurence Checking (Algorithm 6). Consider trace $\sigma = [a_1, a_2, \ldots, a_n]$ and episode with $V = \{v_1, v_2, \ldots, v_m\}$. Worst case, $m = n$.

Finding mapping h is done by, for each v_i find a a_j such that the order condition holds. Checking the order condition takes $O(|\leq|)$. Worst case, we check mappings in ascending order $(v_1 \to a_1, \dots v_1 \to a_n)$ where only the last mapping is valid. Hence, we need $n!$ attempts, resulting in worst case complexity $O(n! \cdot |\leq|)$.

Total Time Complexity of Algorithm 1. The total worst case running time consists of $O(Phase1) + O(Phase2)$, and is given by:

$$O(T_L{}^2 \cdot |\mathcal{A}|^{T_L+1} \cdot \left(|\mathcal{A}|^{T_L+1} + |L| \cdot \Sigma_{l=1}^{T_L}(l-1)! \right)$$
$$+ T_L{}^5 \cdot \Sigma_{l=1}^{\frac{1}{2}T_L{}^2 - \frac{1}{2}T_L} \binom{T_L \cdot (T_L - 1)}{l} \cdot \left(\binom{T_L \cdot (T_L - 1)}{l} + |L| \cdot (T_L - 1)! \right))$$

where: $T_L = \max \{ |\sigma| \mid \sigma \in L \}$ is the max trace size in log, $|L|$ is the size of event log (# trace variants), and $|\mathcal{A}|$ is the size of alphabet (# event classes).

Note that, despite the theoretical worst case time complexity, our episode discovery algorithm is very fast in practice. See also the evaluation in Sect. 5.

4.6   Pruning

Using the pruning techniques described below, we reduce the number of generated episodes (and thereby computation time and memory requirements) and filter out uninteresting results. These techniques eliminate less interesting episodes by ignoring infrequent activities and skipping partial orders on events not occurring relatively close in time. In addition, for pruning based on the antisymmetry of $\leq$ and the Eventually-follows Relation, we leverage the fact that it is cheaper to prune candidates during generation than to eliminate them via occurrence checking.

Activity Pruning. Based on the frequency of an activity, uninteresting episodes can be pruned in an early stage. This is achieved by replacing the activity alphabet $\mathcal{A}$ with the largest set $\mathcal{A}' \subseteq \mathcal{A}$ satisfying $(\forall a \in \mathcal{A}' : ActFreq(a) \geq minActFreq)$, on line 5 in Algorithm 1. This pruning technique allows the episode discovery algorithm to be more resistant to logs with many infrequent activities, which are indicative of exceptions or noise. Note that, if $minActFreq$ is set too high, we can end up with $\mathcal{A}' = \emptyset$. In this case, no episodes are discovered.

Trace Distance Pruning. The pruning of episodes based on a trace distance interval can be achieved by adding the trace distance interval check to line 3 of Algorithm 6. Note that if there are two or more interpretations for h, with one passing and one rejected by the interval check, then we will find the correct interpretation thanks to the $\exists$ on line 7.

Pruning Based on the Antisymmetry of $\leq$. During candidate generation in Algorithm 3 we can leverage the antisymmetry of $\leq$. Recall that in Algorithm 3 we generate candidate episodes γ from merging episodes α and β overlapping on the first $l - 1$ edges. If we extend the predicate on line 9 with the check $reverse(\beta.\leq[l-1]) \notin \alpha.\leq$ we ensure that we don't generate candidate episodes γ that violate the antisymmetry of $\leq$. (Note: $reverse((a, b)) = (b, a)$.)

Pruning Based on the Eventually-Follows Relation. During seeding the partial order candidates in Algorithm 1 on line 15 we can utilize the eventually-follows relation as a fast approximation of early occurrence checking. Using this relation, we can extend the predicate on line 15 with the check $\frac{a \gg_L b}{|L|} \geq minFreq$, where $a = g(v) \wedge b = g(w)$.

In practice, we pre-calculate the eventually-follows matrix, having a space-complexity of $|\mathcal{A}|^2$, where $|\mathcal{A}|$ the number of unique activities in the event log. This allows us to compute the eventually-follows values only once in a linear scan over the log, and reuse values, accessing them in constant time.

4.7   Episode Rule Discovery

The discovery of episode rules is done after discovering all the frequent episodes. For all frequent episodes α, we consider all frequent subepisodes β with $\beta \prec \alpha$ for the episode rule $\beta \Rightarrow \alpha$.

For efficiently finding potential frequent subepisodes β, we use the notion of "discovery tree", based on episode construction. Each time we recognize a frequent episode β created from combining frequent episodes γ and ε, we recognize β as a child of γ and ε. Similarly, γ and ε are the parents of β. See Fig. 3 for an example of a discovery tree.

Using the discovery tree we can walk from an episode α along the discovery parents of α. Each time we find a parent β with $\beta \prec \alpha$, we can consider the parents and children of β. As a result of Lemma 3, we cannot apply pruning in either direction of the parent-child relation based on the confidence $conf(\beta \Rightarrow \alpha)$. This is easy to see for the child direction. For the parent direction, observe the discovery tree in Fig. 3 and $\delta \prec \alpha$. If for episode α we would stop before visiting the parents of β, we would never consider δ (which has $\delta \prec \alpha$).

This principle of traversing the discovery tree is implemented by Algorithm 7. This implementation uses a discovery *front* queue for traversing the discovery tree, similar to the queue used in the Breadth-first search algorithm. The discovery tree is traversed for each discovered episode (each $\alpha \in \Gamma$). Hence, we consider the discovery tree as a partial order on the set Γ, and use that structure to efficiently find sets of subsets.

Algorithm 7. Discovering episode rules

Input: A list of episodes Γ, a confidence threshold $minConf$ and a magnitude interval specified by $minMag$ and $maxMag$.

Output: A set of valid episode rules R

Description: Episode rule discovery. For each discovered episode (each $\alpha \in \Gamma$), the discovery tree is traversed in a Breadth-first search style, searching for candidate β episodes yielding episode rules $\beta \Rightarrow \alpha$.

RULEDISCOVERY($\Gamma, minConf, minMag, maxMag$)
```
(1)      R = ∅
(2)      foreach α ∈ Γ
(3)          discovered = ∅
(4)          Let front be an empty FIFO queue
(5)          foreach parent ∈ α.parents
(6)              discovered = discovered ∪ {parent}
(7)              front.ENQUE(parent)
(8)          while front ≠ ∅
(9)              β = front.DEQUE()
(10)             foreach parent ∈ β.parents
(11)                 discovered = discovered ∪ {parent}
(12)                 front.ENQUE(parent)
(13)             if β ≼ α
(14)                 // prune siblings of α
(15)                 if β ∉ α.parents
(16)                     foreach child ∈ β.children ∧ child ∉ discovered
(17)                         discovered = discovered ∪ {child}
(18)                         front.ENQUE(child)
(19)                     if conf(β ⇒ α) ≥ minConf ∧ minMag ≤ mag(β ⇒ α) ≤
                             maxMag
(20)                         R = R ∪ {β ⇒ α}
```

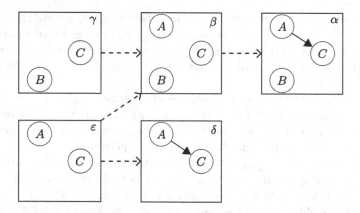

Fig. 3. Part of an example discovery tree. Each block denotes an episode. The dashed arrows between blocks denote a parent-child relationship. In this example we have, amongst others: $\beta \prec \alpha$, $\varepsilon \prec \beta$, $\varepsilon \prec \delta$ and $\delta \prec \alpha$ (not shown as a parent-child relation).

4.8   Implementation Consideration

We implemented the episode discovery algorithm as a ProM 6 plug-in (see also Fig. 7), written in Java. Since the OCCURS() Algorithm 5 is the biggest bottleneck, this part of the implementation was considerably optimized.

5   Evaluation

This section reviews the feasibility of the approach using both synthetic and real-life event data.

5.1   Methodology

We used three different event logs for our experiment. The first event log, *bigger-example.xes*, is an artificial event log from Chap. 5 of [1] and available via http://www.processmining.org/event_logs_and_models_used_in_book. The second and third event logs, *BPI_Challenge_2012.xes* and *BPI_Challenge_2013.xes*, are real life event logs available via doi:10.4121/uuid:3926db30-f712-4394-aebc-75976070e91f and doi:10.4121/uuid:500573e6-accc-4b0c-9576-aa5468b10cee respectively. The experiment consists of two parts: first a series of tests focused on performance and the number of discovered episodes, and second, a case study focused on comparing our technique with existing discovery techniques. For these experiments we used a laptop with a Core i7-4700MQ CPU (2.40 GHz), Java SE Runtime Environment 1.7.0_67 (64 bit) with 4 GB RAM.

5.2   Performance and Number of Discovered Episodes

In Table 3 some key characteristics of the event logs are given. We examined the effects of the parameters *minFreq*, *minActFreq* and *maxTraceDist* on the running time, the discovered number of episodes (number of results), and the total number of intermediate candidate episodes. In Fig. 7 an indication (screenshots) of the ProM plugin output is given.

In Figs. 4, 5, and 6 the results of the experiments are given.

The metric "# Episodes (result)" indicates the size of the end result. This metric is given by $|\Gamma|$ in Algorithm 1. The metric "# Candidate episodes" indicates the size of the intermediate results, after episode construction and pruning, but before occurrence checking. This metric is calculated by summing $|C_l|$ across iterations in both discovery phases in Algorithm 1. The "runtime", indicates the average running time of the algorithm, and its associated 95 % confidence interval. Note that the scale of the runtime is in milliseconds.

Table 3. Metadata for the used event logs.

	# traces	# variants	# activities	events / trace		
				avg.	min.	max.
bigger-example.xes	1,391	21	8	5.42	5	17
BPI_Challenge_2012.xes (BPIC 2012)	13,087	4,366	36	20.05	3	175
BPI_Challenge_2013.xes (BPIC 2013)	7,554	2,278	13	8.68	1	123

As can be seen in the experimental results, we see that the running time is strongly related to the discovered number of episodes. Note that if some parameters are poorly chosen, like high *minFreq* in Fig. 4(b), then a relatively large class of episodes seems to become frequent, thus increasing the running time dramatically.

For a reasonably low number of frequent episodes (<500, more will a human not inspect), the algorithm turns out to be quite fast (under one second). We noted a virtual nonexistent contribution of the parallel episode mining phase to the total running time. This can be explained by a simple combinatorial argument: there are far more partial orders to be considered than there are parallel episodes. Also note the increasing number of candidate episodes in Fig. 5(b), which consists solely of parallel episodes, but there is no significant change in the runtime.

An analysis of the effects of changing the *minFreq* parameter (Fig. 4(a), (b), and (c)) shows that a poorly chosen value results in many episodes. In addition, the *minFreq* parameter gives us fine-grained control of the number of results. It gradually increases the total number of episodes for lower values. Note that, especially for the BPIC 2012 event log, low values for *minFreq* can dramatically increase the running time. This is due to the large number of (candidate) episodes being generated.

Secondly, note that for the *minActFreq* parameter (Fig. 5(a), (b), and (c)), there seems to be a cutoff point that separates frequent from infrequent activities. Small changes around this cutoff point may have a noticeable effect on the number of episodes discovered.

Finally, for the *maxTraceDist* parameter (Fig. 6(a), (b), and (c)), we see that this parameter seems to have a sweet-spot where a low – but not too low – number of episodes are discovered. Chosen a value for *maxTraceDist* just after this sweet-spot yields a large number of episodes.

When comparing the artificial and real life event logs, we see a remarkable pattern. The artificial event log (*bigger-example.xes*), shown in Fig. 4(a) appears to be far more fine-grained than the real life event log (*BPIC 2012*) shown in Fig. 4(b) and (c). In the real life event log there appears to be a clear distinction between frequent and infrequent episodes. In the artificial event log a more fine-grained pattern occurs. Most of the increase in frequent episodes, for decreasing *minFreq*, is again in the partial order discovery phase.

5.3   Case Study – Pattern Discovery Compared with Existing Algorithms

As noted in the introduction, often the overall end-to-end process models are rather complicated. Therefore, the search for local patterns (i.e., episodes) is interesting. In this section we perform a short case study using the BPI Challenge 2012, an event log of a loan application process. We explored this event log using: the α-algorithm [11], Heuristics miner [14], Inductive miner [22], DECLARE

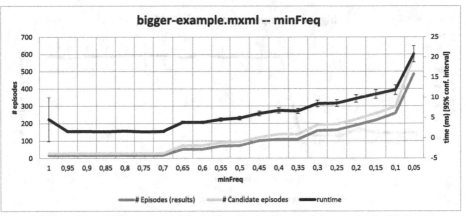

(a) Event log: bigger-example.mxml , $minActFreq = 1.0$, $maxTraceDist = 4$

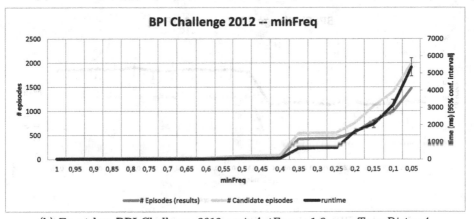

(b) Event log: BPI Challenge 2012 , $minActFreq = 1.0$, $maxTraceDist = 4$

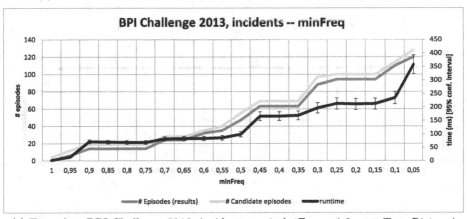

(c) Event log: BPI Challenge 2013, incidents , $minActFreq = 1.0$, $maxTraceDist = 4$

Fig. 4. Effects of the parameter *minFreq* on the number of results and candidate episodes. Observe that the *minFreq* parameter gives us fine-grained control of the number of results. Note that for less than 500 result episodes, the runtime is less than one second.

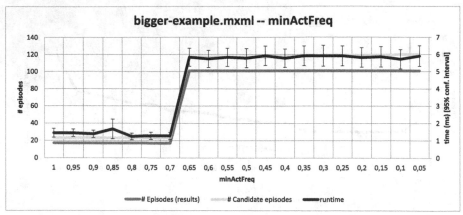

(a) Event log: bigger-example.mxml , $minFreq = 0.45$, $maxTraceDist = 4$

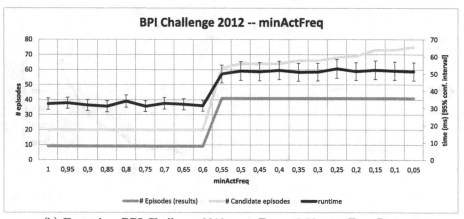

(b) Event log: BPI Challenge 2012 , $minFreq = 0.50$, $maxTraceDist = 4$

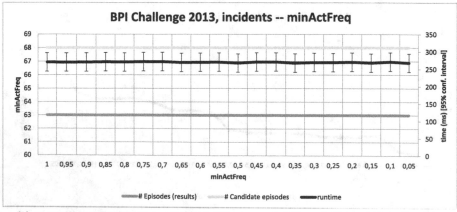

(c) Event log: BPI Challenge 2013, incidents , $minFreq = 0.45$, $maxTraceDist = 4$

Fig. 5. Effects of the parameter $minActFreq$ on the number of results and candidate episodes. Observe that there seems to be a cutoff point that separates frequent from infrequent activities. Note that the runtime is never greater than a third of a second.

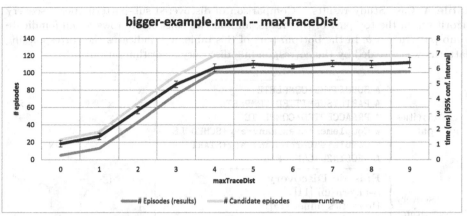

(a) Event log: bigger-example.mxml , $minFreq = 0.45$, $minActFreq = 0.65$

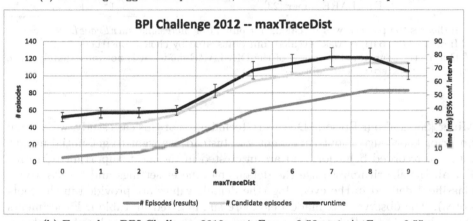

(b) Event log: BPI Challenge 2012 , $minFreq = 0.50$, $minActFreq = 0.55$

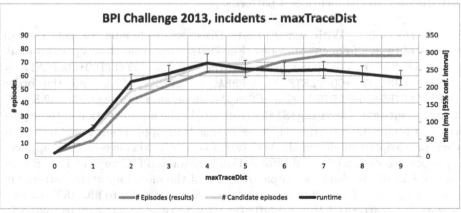

(c) Event log: BPI Challenge 2013, incidents , $minFreq = 0.45$, $minActFreq = 1.00$

Fig. 6. Effects of the parameter $maxTraceDist$ on the number of results and candidate episodes. Observe that $maxTraceDist$ seems to have a sweet-spot where a low – but not too low – number of episodes are discovered. Note that the runtime is never greater than a third of a second.

Table 4. Case Study results – Comparison of discovered sub-patterns per discovery algorithm. In the top part of this table, an x in two consecutive rows a and b indicate a sub-pattern $a \leq b$. In the bottom part of this table, a + indicates the corresponding patterns was revealed by the corresponding discovery algorithm output.

Activities and pattern	A_SUBMITTED+COMPLETE	x				
	A_PARTLYSUBMITTED+COMPLETE	x	x			x
	A_PREACCEPTED+COMPLETE		x	x		
	W_Complementeren_aanvraag+SCHEDULE			x	x	
	W_Complementeren_aanvraag+START				x	
	A_DECLINED+COMPLETE					x
Discovery algorithms	**Episode Discovery**	+	+	+	+	+[a]
	α-algorithm [11]	+				
	Heuristics miner [14]	+		+	+	
	Inductive miner [22]	+	+[b]	+[b]	+	+[b]
	DECLARE miner [23]	+	+[c]	+	+	+[c]

[a] Indicates the pattern was revealed, but only after increasing $maxTraceDist$.

[b] Indicates the pattern was revealed, but obfuscated by choice constructs.

[c] Due to the aggregated overview of the DECLARE model, it is not immediately clear that these patterns are disjoint.

Miner [23], and our Episode Discovery technique. For this case study, we assume no prior knowledge about this event log. Instead, we want to get initial insight into the recorded behavior, and are interested in the most important patterns. For all the algorithms we use the default parameter settings and the "Activity classifier" defined in the event log (the default values are provided in the footnotes). The observations made below are summarized in Table 4. Experiments show that only the Episode Discovery was able to unobfuscated and unambiguously discover all the mentioned patterns.

Episode Discovery. With our Episode Discovery technique we get a small overview of twelve frequent episodes (Fig. 7(a)). Inspecting these episodes more closely, we find two frequent patterns: the order A_SUBMITTED+COMPLETE $\leq$ A_PARTLYSUBMITTED+COMPLETE $\leq$ A_PREACCEPTED+COMPLETE, and the order A_PREACCEPTED+COMPLETE $\leq$ W_Complementeren_aanvraag+SCHEDULE $\leq$ W_Complementeren_aanvraag+START (Fig. 7(b)). The interpretation of these patterns is twofold. One, frequently whenever a loan application is submitted it either preaccepted or declined. And two, frequently whenever a loan application is preaccepted, additional information is requested ("Complementeren aanvraag"). Clearly, we found a simple overview of the most important patterns in the event log. After increasing the $maxTraceDist$ parameter to fifty (50), we also discover the patternA_PARTLYSUBMITTED+COMPLETE $\leq$ A_DECLINED+COMPLETE (see Fig. 7(c)). In the remaining paragraph, we focus on finding patterns using other discovery techniques, and we are particularly interested in finding similar patterns.

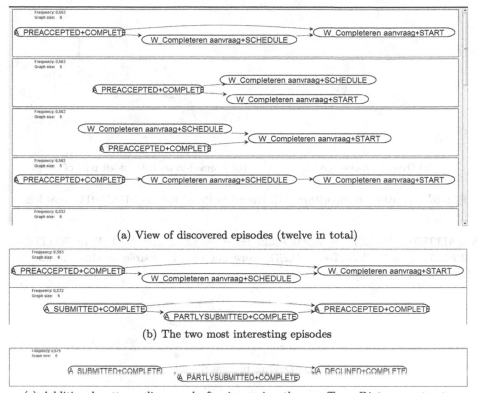

(a) View of discovered episodes (twelve in total)

(b) The two most interesting episodes

(c) Additional pattern, discovered after increasing the *maxTraceDist* parameter to fifty (50).

Fig. 7. Algorithm: Episode Discovery. Result in ProM for the BPIC 2012 event log.

α-algorithm.[1] Figure 8(a) shows the overall Petri net model produced by the α-algorithm [11]. Closer inspection of the bottom-left part (Fig. 8(b)) reveals the sub-pattern A_SUBMITTED+COMPLETE ≤ A_PARTLYSUBMITTED+COMPLETE. The remaining of the previously discovered frequent patterns are not clearly visible in this model. No other patterns were discovered.

Heurisitcs miner.[2] The heuristics net in Fig. 9(a) is produced by the Heuristics miner [14]. Closer inspection of this net (Fig. 9(b)) reveals two sub-patterns: the order A_SUBMITTED+COMPLETE ≤ A_PARTLYSUBMITTED+COMPLETE, and the order A_PREACCEPTED+COMPLETE ≤ W_Complementeren_aanvraag+SCHEDULE ≤ W_Complementeren_aanvraag+START. However, the sub-pattern A_PARTLY

---

[1] **Plugin action:** "Mine for a Petri Net using Alpha-algorithm".
   Parameters: n/a.
[2] **Plugin action:** "Mine for a Heuristics Net using Heuristics Miner".
   Parameters: Activity classifier, Relative-to-best = 5.0, Dependency = 90.0, Length-one-loops = 90.0, Length-two-loops = 90.0, Long distance = 90.0, All tasks connected = On, Long distance dependency = Off, Ignore loop dependency tresholds = On.

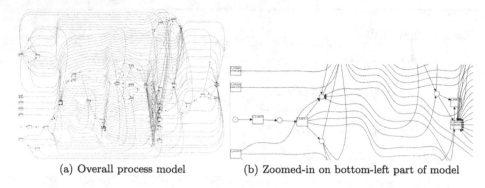

(a) Overall process model   (b) Zoomed-in on bottom-left part of model

Fig. 8. Algorithm: α-algorithm [11]. Result in ProM for the BPIC 2012 event log.

SUBMITTED+ COMPLETE $\leq$ A_PREACCEPTED+COMPLETE and A_PARTLYSUBMITTED+ COMPLETE $\leq$ A_DECLINED+COMPLETE were not clearly visible in this model. No other patterns were discovered.

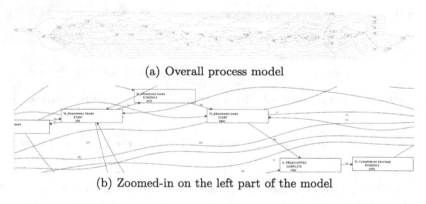

(a) Overall process model

(b) Zoomed-in on the left part of the model

Fig. 9. Algorithm: Heuristics miner [14]. Result in ProM for the BPIC 2012 event log.

Inductive miner.[3] Fig. 10(a) shows the overall process model (a process tree) produced by the Inductive miner [22]. All frequent patterns can be found in this model. However, as can be seen in the close-up in Fig. 10(b), the choice constructs obfuscate these patterns. After detailed inspection of this model, and armed with our results from the Episode Discovery technique, we discovered one less frequent pattern. We rephrase our first interpretation of the Episode Discovery results as: "whenever a loan application is submitted it frequently either preaccepted or declined, or in some rare cases followed by a fraud detection" ("Beoordelen fraude").

---

[3] **Plugin action:** "Mine process tree with Inductive Miner".
   Parameters: Variant = Inductive Miner - infrequent, Noise threshold = 0.20, Event classifier = Event Name.

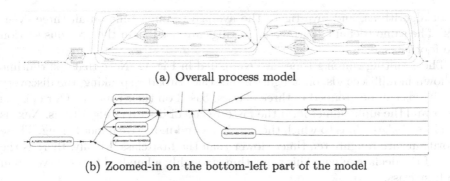

(a) Overall process model

(b) Zoomed-in on the bottom-left part of the model

Fig. 10. Algorithm: Inductive miner [22]. Result in ProM for the BPIC 2012 event log.

DECLARE Miner.[4] Finally, in Fig. 11, the DECLARE model is given, as produced by the DECLARE Miner [23]. In this case we did change the following
parameters: we chose the *succession* template and set the min support to 50
(comparable to the default settings of Episode Miner). As can be observed,
all the frequent patterns can be found. However, note that due to the aggregated overview of the DECLARE model, it is not immediately clear that
the patterns A_PARTLYSUBMITTED+COMPLETE $\leq$ A_PREACCEPTED+COMPLETE and
A_PARTLYSUBMITTED+COMPLETE $\leq$ A_DECLINED+COMPLETE are disjoint. No other
patterns were discovered.

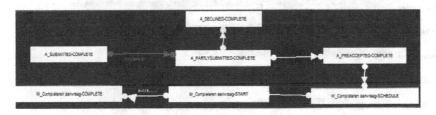

Fig. 11. Algorithm: DECLARE Miner [23]. Result in ProM for the BPIC 2012 event
log, using the *succession* template and a min support of 50.

As demonstrated in this case study, and summarized in Table 4, overall end-
to-end process models can be rather complicated, and the search for local patterns (i.e., episodes) quickly reveals important insight into recorded behavior.

5.4   Case Study – Runtime Compared with Existing Algorithms

After showing the insights that can be gained by our algorithm, we now compare
the running time of our approach with existing algorithms. We revisit the same

---

[4] **Plugin action:** "Declare Maps Miner".
Parameters: Selected Templates = {succession}, All Activities (considering Event Types), Min.
support = 50, Alpha = 0, Control_Flow = On, Time = Off.

set of algorithms, and investigate the average running time on all three event logs. The same (default) parameter settings are used as in the previous section (see footnotes 1-4).

The resulting running times are compared in Fig. 12. Note that the runtime is shown in milliseconds, on a logarithmic scale. Broadly speaking, the discovery algorithms can be grouped in three classes, based on their runtime. Our episode miner and the alpha miner form the fastest class of discovery algorithms. Next is the class of algorithms to which the heuristics and inductive miner belong. These algorithms are roughly ten times slower than the first class. Finally, there is the class of the declare miner. This algorithm is roughly a hundred times slower than the first class.

Looking at the difference between the BPIC 2012 and 2013 logs, we see observe the 2012 log has more event classes (36 for 2012, 13 for 2013), more traces (13,087 for 2012, 7,554 for 2013), and longer traces (avg. 20.05 for 2012, avg. 8.68 for 2013). This increase in size is directly observable in terms of running time for the existing algorithms, but has a less effect on the running time of the episode miner (with default settings).

We conclude that our Episode Discovery realization is among the fastest of algorithms. In particular, it is orders of magnitude faster than the Declare Miner configured to discover only succession relations.

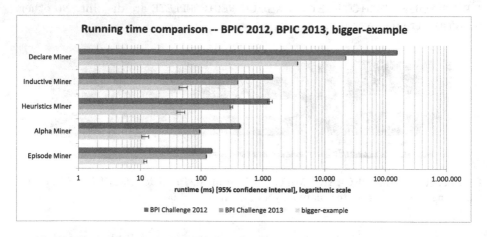

Fig. 12. Comparison of the running time for the different discovery algorithms used in the case study. The runtime is shown in milliseconds, on a logarithmic scale. We distinguish three classes based on runtimes: 1) our Episode miner and the α-miner, 2) the class of algorithms to which the Heuristics and Inductive miner belong, and 3) the class of the Declare miner.

5.5   Case Study – Episode Rules

Continuing with our case study of the BPI Challenge 2012 event log, we also take a look at the discovery of association rules. Here we use the episode rule generation feature of our Episode Discovery ProM plugin, and used the default settings.

The result consists of six episode rules, one of which is shown in Fig. 13. The interpretation of the shown episode rule is as follows: "If we saw A_PARTLY SUBMITTED+COMPLETE ≤ A_PREACCEPTED+COMPLETE occurring, we likely will also see W_Complementeren_aanvraag+SCHEDULE occurring next". In other words, whenever a partially submitted request was preaccepted, it is likely that we will request additional information ("Complementeren aanvraag").

Similar, episode rules can be used in an online setting to predict likely follow-up activities using episodes discovered in historical data.

Fig. 13. Episode rules discovered in ProM for the BPIC 2012 event log. The black solid line indicates the assumed partial order (the β in $\beta \Rightarrow \alpha$), the red dashed line indicates the added pattern (the α) (Color figure online.)

5.6 Case Study – Alternative Perspective: Resources

We conclude our case study of the BPI Challenge 2012 event log with mining patterns in the flow of work between persons. For this we used the Resource classifier defined in the event log. We explored this perspective using: the Inductive miner [22], Handover of Work Social Network miner [26], and our Episode Discovery technique.

The discovered episodes are shown in Fig. 14. The vertices in these results represent resources instead of activities. The first pattern shows that the resource 112 is present in all traces (based on the observation that $freq(112 \leq 112 \leq 112) = 1.0$). Furthermore, we also discover that in most cases work is passed from the resource 112 to tasks without a recorded resource (e.g., automated tasks). Activities conducted by "no recorded resource" can be observed in Fig. 14 as empty vertices.

Figure 15(a) shows the overall process model (a process tree) for the resource perspective, produced by the Inductive miner [22]. At first glance no obvious pattern is visible. In the close-up in Fig. 15(b), the resource 112 and "no recorded resource"/"empty resource" are visible, but no clear patterns are visible.

In Fig. 15(c) the handover of work social network is given, as produced by the organizational miner [26]. Most of the resources are forming one big tightly-connected cluster. The "no recorded resource"/"empty resource" is completely disconnected, but the resource 112 is not easily found (it is in the top-left corner). The patterns found by the Episode Miner cannot be deduced from this social network.

By using the resource perspective in combination with Episode Discovery, we gained insight into the most important resources, and the flow of work between

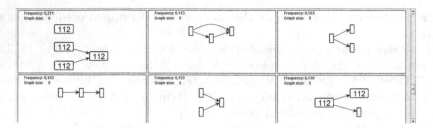

Fig. 14. Episodes discovered in ProM for the BPIC 2012 event log, using the Resource classifier. In total, forty episodes were discovered. Note that the vertices in these results represent resources instead of activities. The empty vertices indicate the absence of a recorded resource (e.g., automated tasks).

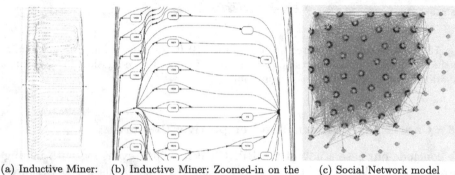

(a) Inductive Miner: (b) Inductive Miner: Zoomed-in on the   (c) Social Network model
Overall process model   top part of the model

Fig. 15. Result in ProM for the BPIC 2012 event log, using the Resource classifier. Algorithms: Inductive miner [22], Handover of Work Social Network miner [26].

resources. This demonstrates that Episode Discovery is not only useful in the activity-focused control-flow perspective, but also in other perspectives. While we only showed pattern discovery in the control-flow and resource domain, other perspectives are possible. One example is discovering the flow of work between event locations (e.g., system components or organization departments generating the events). Another example is discovering the relations between data attributes (e.g., which information is used in which order).

6 Conclusion and Future Work

In this paper, we considered the problem of discovering frequently occurring episodes in an event log. An episode is a collection of events that occur in a given partial order. We presented efficient algorithms for the discovery of frequent episodes and episode rules occurring in an event log, and presented experimental results.

Our experimental evaluation shows that, for a reasonably low number of frequent episodes, the algorithm turns out to be quite fast (under one second);

typically faster than existing many algorithms. The main problem is the correct setting of the episode pruning parameters $minFreq$, $minActFreq$, and $maxTraceDist$. In addition, comparison with existing discovery algorithms has shown the benefit of episode mining in getting insight into recorded behavior. Moreover, we have demonstrated the usefulness of episode rules that can be discovered. Finally, the applicability of Episode Discovery for other perspectives (like the resources perspective) was shown.

During the development of the algorithm for ProM 6, special attention was paid to optimizing the OCCURS() algorithm (Algorithm 5) implementation, which proved to be the main bottleneck. Future work could be to prune occurrence checking based on the parents of an episode, leveraging the fact that an episode cannot occur in a trace if a parent also did occur in that trace.

Another approach to improve the algorithm is to apply the *generic divide and conquer approach for process mining*, as defined in [28]. This approach splits the set of activities into a collection of partly overlapping activity sets. For each activity set, the log is projected onto the relevant events, and the regular episode discovery algorithm is applied. In essence, the same trick is applied as used by the $minActFreq$ parameter (using an alphabet subset), which is to create a different set of initial 1-node parallel episodes to start discovering with.

The main bottleneck is the frequency computation by checking the occurrence of each episode in each trace. Typically, we have a small amount of episodes to check, but many traces to check against. Using the MapReduce programming model developed by Dean and Ghemawat, we can easily parallelize the episode discovery algorithm and execute it on a large cluster of commodity machines [29]. The MapReduce programming model requires us to define *map* and *reduce* functions. The *map* function, in our case, accepts a trace and produces [episode, trace] pairs for each episode occurring in the given trace. The *reduce* function accepts an episode plus a list of traces in which that episode occurs, and outputs a singleton list if the episode is frequent, and an empty list otherwise. This way, the main bottleneck of the algorithm can be effectively parallelized.

References

1. van der Aalst, W.M.P.: Process Mining: Discovery, Conformance and Enhancement of Business Processes. Springer, Berlin (2011)
2. Mannila, H., Toivonen, H., Verkamo, A.I.: Discovery of frequent episodes in event sequences. Data Min. Knowl. Disc. 1(3), 259–289 (1997)
3. Lu, X., Fahland, D., van der Aalst, W.M.P.: Conformance checking based on partially ordered event data. To appear in Business Process Intelligence 2014, workshop SBS (2014)
4. Agrawal, R., Srikant, R.: Fast algorithms for mining association rules in large databases. In: Proceedings of the 20th International Conference on Very Large Data Bases, VLDB 1994, pp. 487–499. Morgan Kaufmann Publishers Inc., San Francisco (1994)
5. Agrawal, R., Srikant, R.: Mining sequential patterns. In: Proceedings of the Eleventh International Conference on Data Engineering, ICDE 1995, pp. 3–14. IEEE Computer Society, Washington, D.C. (1995)

6. Srikant, R., Agrawal, R.: Mining sequential patterns: generalizations and perfor-mance improvements. In: Apers, P., Bouzeghoub, M., Gardarin, G. (eds.) EDBT 1996. LNCS, vol. 1057, pp. 1–17. Springer, Heidelberg (1996)
7. Lu, X., Mans, R.S., Fahland, D., van der Aalst, W.M.P.: Conformance checking in healthcare based on partially ordered event data. In: Grau, A., Zurawski, R. (eds.) ETFA 2014, pp. 1–8. IEEE Computer Society, Barcelona (2014)
8. Fahland, D., van der Aalst, W.M.P.: Model repair: aligning process models to reality. Inf. Syst. **47**, 220–243 (2015)
9. Leemans, M.: Episode miner. https://svn.win.tue.nl/repos/prom/Packages/ EpisodeMiner/. Accessed 9 January 2015
10. Laxman, S., Sastry, P.S., Unnikrishnan, K.P.: Fast algorithms for frequent episode discovery in event sequences. In: Proceedings of the 3rd workshop on mining tem-poral and sequential data, SIGKDD, Seattle, WA, USA. Association for Computing Machinery, Inc., August 2004
11. van der Aalst, W.M.P., Weijters, A.J.M.M., Maruster, L.: Workflow mining: dis-covering process models from event logs. IEEE Trans. Knowl. Data Eng. **16**(9), 1128–1142 (2004)
12. de Medeiros, A.K.A., van der Aalst, W.M.P., Weijters, A.J.M.M.T.: Workflow mining: current status and future directions. In: Meersman, R., Schmidt, D.C. (eds.) CoopIS 2003, DOA 2003, and ODBASE 2003. LNCS, vol. 2888, pp. 389–406. Springer, Heidelberg (2003)
13. Wen, L., van der Aalst, W.M.P., Wang, J., Sun, J.: Mining process models with non-free-choice constructs. Data Min. Knowl. Disc. **15**(2), 145–180 (2007)
14. Weijters, A.J.M.M., van der Aalst, W.M.P., de Medeiros, A.K.A.: Process mining with the heuristics miner-algorithm. In: BETA Working Paper Series, WP 166. Eindhoven University of Technology, Eindhoven (2006)
15. de Medeiros, A.K.A., Weijters, A.J.M.M., van der Aalst, W.M.P.: Genetic process mining: an experimental evaluation. Data Min. Knowl. Disc. **14**(2), 245–304 (2007)
16. Buijs, J.C.A.M., van Dongen, B.F., van der Aalst, W.M.P.: On the role of fitness, precision, generalization and simplicity in process discovery. In: Meersman, R., Panetto, H., Dillon, T., Rinderle-Ma, S., Dadam, P., Zhou, X., Pearson, S., Ferscha, A., Bergamaschi, S., Cruz, I.F. (eds.) OTM 2012, Part I. LNCS, vol. 7565, pp. 305–322. Springer, Heidelberg (2012)
17. Solé, M., Carmona, J.: Process mining from a basis of state regions. In: Lilius, J., Penczek, W. (eds.) PETRI NETS 2010. LNCS, vol. 6128, pp. 226–245. Springer, Heidelberg (2010)
18. van der Aalst, W.M.P., Rubin, V., Verbeek, H.M.W., van Dongen, B.F., Kindler, E., Günther, C.W.: Process mining: a two-step approach to balance between under-fitting and overfitting. Softw. Syst. Model. **9**(1), 87–111 (2010)
19. Bergenthum, R., Desel, J., Lorenz, R., Mauser, S.: Process mining based on regions of languages. In: Alonso, G., Dadam, P., Rosemann, M. (eds.) BPM 2007. LNCS, vol. 4714, pp. 375–383. Springer, Heidelberg (2007)
20. van der Werf, J.M.E.M., van Dongen, B.F., Hurkens, C.A.J., Serebrenik, A.: Process discovery using integer linear programming. Fundamenta Informaticae **94**, 387–412 (2010)
21. Günther, C.W., van der Aalst, W.M.P.: Fuzzy mining – adaptive process simplifi-cation based on multi-perspective metrics. In: Alonso, G., Dadam, P., Rosemann, M. (eds.) BPM 2007. LNCS, vol. 4714, pp. 328–343. Springer, Heidelberg (2007)
22. Leemans, S.J.J., Fahland, D., van der Aalst, W.M.P.: Discovering block-structured process models from incomplete event logs. In: Ciardo, G., Kindler, E. (eds.) PETRI NETS 2014. LNCS, vol. 8489, pp. 91–110. Springer, Heidelberg (2014)

23. Maggi, F.M., Mooij, A.J., van der Aalst, W.M.P.: User-guided discovery of declarative process models. In: 2011 IEEE Symposium on Computational Intelligence and Data Mining (CIDM), pp. 192–199. IEEE (2011)
24. Maggi, F.M., Bose, R.P.J.C., van der Aalst, W.M.P.: Efficient discovery of understandable declarative process models from event logs. In: Ralyté, J., Franch, X., Brinkkemper, S., Wrycza, S. (eds.) CAiSE 2012. LNCS, vol. 7328, pp. 270–285. Springer, Heidelberg (2012)
25. Maggi, F.M., Bose, R.P.J.C., van der Aalst, W.M.P.: A knowledge-based integrated approach for discovering and repairing declare maps. In: Salinesi, C., Norrie, M.C., Pastor, Ó. (eds.) CAiSE 2013. LNCS, vol. 7908, pp. 433–448. Springer, Heidelberg (2013)
26. Song, M., van der Aalst, W.M.P.: Towards comprehensive support for organizational mining. Decis. Support Syst. 46(1), 300–317 (2008)
27. Kryszkiewicz, M.: Fast discovery of representative association rules. In: Polkowski, L., Skowron, A. (eds.) RSCTC 1998. LNCS (LNAI), vol. 1424, p. 214. Springer, Heidelberg (1998)
28. van der Aalst, W.M.P.: Decomposing petri nets for process mining: a generic approach. Distrib. Parallel Databases 31(4), 471–507 (2013)
29. Dean, J., Ghemawat, S.: MapReduce: simplified data processing on large clusters. Commun. ACM 51(1), 107–113 (2008)

Finding Suitable Activity Clusters
for Decomposed Process Discovery

B.F.A. Hompes[✉], H.M.W. Verbeek, and W.M.P. van der Aalst

Department of Mathematics and Computer Science,
Eindhoven University of Technology, Eindhoven, The Netherlands
{b.f.a.hompes,h.m.w.verbeek,w.m.p.v.d.aalst}@tue.nl

Abstract. Event data can be found in any information system and provide the starting point for a range of process mining techniques. The widespread availability of large amounts of event data also creates new challenges. Existing process mining techniques are often unable to handle "big event data" adequately. *Decomposed process mining* aims to solve this problem by decomposing the process mining problem into many smaller problems which can be solved in less time, using less resources, or even in parallel. Many decomposed process mining techniques have been proposed in literature. Analysis shows that even though the decomposition step takes a relatively small amount of time, it is of key importance in finding a high-quality process model and for the computation time required to discover the individual parts. Currently there is no way to assess the quality of a decomposition beforehand. We define three quality notions that can be used to assess a decomposition, before using it to discover a model or check conformance with. We then propose a decomposition approach that uses these notions and is able to find a high-quality decomposition in little time.

Keywords: Decomposed process mining · Decomposed process discovery · Distributed computing · Event log

1 Introduction

Process mining aims to discover, monitor and improve real processes by extracting knowledge from event logs readily available in today's information systems [1]. In recent years, (business) processes have seen an explosive rise in supporting infrastructure, information systems and recorded information, as illustrated by the term Big Data. As a result, event logs generated by these information systems grow bigger and bigger as more event (meta-)data is being recorded and processes grow in complexity. This poses both opportunities and challenges for the process mining field, as more knowledge can be extracted from the recorded data, increasing the practical relevance and potential economic value of process mining. Traditional process mining approaches however have difficulties coping with this sheer amount of data (i.e. the number of events), as

© IFIP International Federation for Information Processing 2015
P. Ceravolo et al. (Eds.): SIMPDA 2014, LNBIP 237, pp. 32–57, 2015.
DOI: 10.1007/978-3-319-27243-6_2

most interesting algorithms are linear in the size of the event log and exponential in the number of different activities [2].

In order to provide a solution to this problem, techniques for *decomposed process mining* [2–4] have been proposed. Decomposed process mining aims to decompose the process mining problem at hand into smaller problems that can be handled by existing process discovery and conformance checking techniques. The results for these individual sub-problems can then be combined into solutions for the original problems. Also, these smaller problems can be solved concurrently with the use of parallel computing. Even sequentially solving many smaller problems can be faster than solving one big problem, due to the exponential nature of many process mining algorithms [2]. Several decomposed process mining techniques have been developed in recent years [2–4, 9, 10, 12, 16]. Though existing approaches have their merits, they lack in generality. In [3], a generic approach to decomposed process mining is proposed. The proposed approach provides a framework which can be combined with different existing process discovery and conformance checking techniques. Moreover, different decompositions can be used while still providing formal guarantees, e.g. the fraction of perfectly fitting traces is not influenced by the decomposition.

When decomposing an event log for (decomposed) process mining, several problems arise. In terms of decomposed process discovery, these problems lie in the step where the overall event log is decomposed into sublogs, where submodels are discovered from these sublogs, and/or where submodels are merged to form the final model. Even though creating a decomposition is computationally undemanding, it is of key importance for the remainder of the decomposed process discovery process in terms of the overall required processing time and the quality of the resulting process model.

The problem is that there is currently no clear way of determining the quality of a given decomposition of the events in an event log, before using that decomposition to either discover a process model or check conformance with.

The current decomposition approaches do not use any quality notions to create a decomposition. Thus, potential improvements lie in finding such quality notions and a decomposition approach that uses those notions to create a decomposition with.

The remainder of this paper is organized as follows. In Sect. 2 related work is discussed. Section 3 introduces necessary preliminary definitions for decomposed process mining and the generic decomposition approach. Section 4 introduces decomposition quality notions to grade a decomposition upon, and two approaches that create a high quality decomposition according to those notions. Section 5 shows a (small) use case. The paper is concluded with views on future work in Sect. 6.

2   Related Work

Process discovery aims at discovering a process model from an event log while conformance checking aims at diagnosing the differences between observed and modeled behavior (resp. the event log and the model). Various discovery algorithms and many different modeling formalisms have been proposed in literature. As very different approaches are used, it is impossible to provide a complete overview of all techniques here. We refer to [1] for an introduction to process mining and an overview of existing techniques. For an overview of best practices and challenges, we refer to the Process Mining Manifesto [5]. The goal of this paper is to improve decomposed process discovery, where challenging discovery problems are split in many smaller problems which can be solved by existing discovery techniques.

In the fields of data mining and machine learning many efforts have been made to improve the scalability of existing techniques. Most of these techniques can be distributed [8,17], e.g. distributed clustering, distributed classification, and distributed association rule mining. To support this, several distributed data processing platforms have been created and are widely used [14,20]. Some examples are Apache Hadoop [23], Spark [26], Flink [21], and Tez [19]. Specific data mining and machine learning libraries are available for most of these platforms. However, these approaches often partition the input data and therefore cannot be used for the discovery of process models. Decomposed process mining aims to provide a solution to this problem.

Little work has been done on the decomposition and distribution of process mining problems [2–4]. In [18] MapReduce is used to scale event correlation as a preprocessing step for process mining. More related are graph-partitioning based approaches. By partitioning a causal dependency graph into partially overlapping subgraphs, events are clustered into groups of events that are causally related. In [11] it is shown that region-based synthesis can be done at the level of synchronized State Machine Components (SMCs). Also a heuristic is given to partition the causal dependency graph into overlapping sets of events that are used to construct sets of SMCs. Other region-based decomposition techniques are proposed in [9,10]. However, these techniques are limited to discovering Petri Nets from event logs. In [16] the notions of Single-Entry Single-Exit (SESE) and Refined Process Structure Trees (RPSTs) are used to hierarchically partition a process model and/or event log for decomposed process discovery and conformance checking. In [6], passages are used to decompose process mining problems.

In [3] a different (more local) partitioning of the problem is given which, unlike [11], decouples the decomposition approach from the actual conformance checking and process discovery approaches. It is indicated that a partitioning of the activities in the event log can be made based on a causal graph of activities. It can therefore be used together with any of the existing process discovery and conformance checking techniques. The approach presented in this paper is an extension of the approach presented in [3], though we focus on discovery. Where [3] splits the process mining problem at hand into subproblems using a *maximal* decomposition of a causal dependency graph, our approach first aims to recombine the many created activity clusters into better and fewer clusters, and only then splits the process mining problem into subproblems. As a result, fewer subproblems remain to be solved.

The techniques used to recombine clusters are inspired by existing, well-known software quality metrics and the business process metrics listed in [22], i.e. *cohesion* and *coupling*. More information on the use of software engineering metrics in a process mining context is described in [22]. However, in this instance, these metrics are used to measure the quality of the decomposition itself rather than the process to be discovered. As such, the quality of the decomposition can be assessed before it is used to distribute the process mining problem.

3   Preliminaries

This section introduces the notations needed to define a better decomposition approach. A basic understanding of process mining is assumed [1].

3.1   Multisets, Functions, and Sequences

Definition 1 (Multisets). *Multisets are defined as sets where elements may appear multiple times.* $\mathcal{B}(A)$ *is the set of all multisets over some set A. For some multiset $b \in \mathcal{B}(A)$, and element $a \in A$, $b(a)$ denotes the number of times a appears in b.*

For example, take $A = \{a, b, c, d\} : b_1 = []$ denotes the empty multiset, $b_2 = [a, b]$ denotes the multiset over A where $b_2(c) = b_2(d) = 0$ and $b_2(a) = b_2(b) = 1, b_3 = [a, b, c, d]$ denotes the multiset over A where $b_3(a) = b_3(b) = b_3(c) = b_3(d) = 1, b_4 = [a, b, b, d, a, c]$ denotes the multiset over A where $b_4(a) = b_4(b) = 2$ and $b_4(c) = b_4(d) = 1$, and $b_5 = [a^2, b^2, c, d] = b_4$. The standard set operators can be extended to multisets, e.g. $a \in b_2, b_5 \setminus b_2 = b_3, b_2 \uplus b_3 = b_4 = b_5, |b_5| = 6$

Definition 2 (Function Projection). *Let $f \in X \nrightarrow Y$ be a (partial) function and $Q \subseteq X$. $f\!\restriction_Q$ denotes the projection of f on $Q : dom(f\!\restriction_Q) = dom(f) \cap Q$ and $f\!\restriction_Q(x) = f(x)$ for $x \in dom(f\!\restriction_Q)$.*

The projection can be used for multisets. For example, $b_5\!\restriction_{\{a,b\}} = [a^2, b^2]$.

Definition 3 (Sequences). *A sequence is defined as an ordering of elements of some set. Sequences are used to represent paths in a graph and traces in an event log. $\mathcal{S}(A)$ is the set of all sequences over some set A. $s = \langle a_1, a_2, \ldots, a_n \rangle \in \mathcal{S}(A)$ denotes a sequence s over A of length n. Furthermore: $s_1 = \langle \, \rangle$ is the empty sequence and $s_1 \cdot s_2$ is the concatenation of two sequences.*

For example, take $A = \{a, b, c, d\} : s_1 = \langle a, b, b \rangle, s_2 = \langle b, b, c, d \rangle, s_1 \cdot s_2 = \langle a, b, b, b, b, c, d \rangle$

Definition 4 (Sequence Projection). *Let A be a set and $Q \subseteq A$ a subset. $\restriction_Q \in \mathcal{S}(A) \rightarrow \mathcal{S}(Q)$ is a projection function and is defined recursively: (1) $\langle \, \rangle\!\restriction_Q = \langle \, \rangle$ and (2) for $s \in \mathcal{S}(A)$ and $a \in A$:*

$$(\langle a \rangle \cdot s)\!\restriction_Q = \begin{cases} s\!\restriction_Q & \text{if } a \notin Q \\ \langle a \rangle \cdot s\!\restriction_Q & \text{if } a \in Q \end{cases}$$

So $\langle a, a, b, b, c, d, d \rangle\!\restriction_{\{a,b\}} = \langle a, a, b, b \rangle$.

3.2   Event Logs

Event logs are the starting point for process mining. They contain information recorded by the information systems and resources supporting a process. Typically, the executed *activities* of multiple *cases* of a *process* are recorded. Note that only *example behavior* is recorded, i.e. event logs only contain information that has been seen. An event log often contains only a fraction of the possible behavior [1]. A trace describes one specific instance (i.e. one case) of the process at hand, in terms of the executed activities. An event log is a multiset of traces, since there can be multiple cases having the same trace. For the remainder of this paper, we let $\mathcal{U}_A$ be some universe of activities.

Definition 5 (Event log). *Let $A \subseteq \mathcal{U}_A$ be a set of activities. A trace $s \in \mathcal{S}(A)$ is a sequence of activities. Let $L \in \mathcal{B}(\mathcal{S}(A))$ be a multiset of traces over A. L is an event log over A.*

An example event log is $L_1 = [\langle a, b, c, d \rangle^5, \langle a, b, b, c, d \rangle^2, \langle a, c, d \rangle^3]$. There are three unique traces in L_1, and it contains information about a total of 10 cases. There are $4 \cdot 5 + 5 \cdot 2 + 3 \cdot 3 = 39$ events in total. The projection can be used for event logs as well. That is, for some log $L \in \mathcal{B}(\mathcal{S}(A))$ and set $Q \subseteq A : L \lceil_Q = [s \lceil_Q | s \in L]$. For example $L_1 \lceil_{\{a,b,c\}} = [\langle a, b, c \rangle^5, \langle a, b, b, c \rangle^2, \langle a, c \rangle^3]$. We will refer to these projected event logs as *sublogs*.

3.3   Activity Matrices, Graphs, and Clusters

In [3] different steps for a generic decomposed process mining approach have been outlined. We *decompose* the overall event log based on a *causal graph* of activities. This section describes the necessary definitions for this decomposition method.

Definition 6 (Causal Activity Matrix). *Let $A \subseteq \mathcal{U}_A$ be a set of activities. $\mathcal{M}(A) = (A \times A) \rightarrow [-1.0, 1.0]$ denotes the set of causal activity matrices over A. For $a_1, a_2 \in A$ and $M \in \mathcal{M}(A)$, $M(a_1, a_2)$ denotes the "causal relation strength" from a_1 to a_2.*

A value close to 1.0 signifies that we are quite confident there exists a causal relation (e.g. directly follows relation) between two activities while a value close to -1.0 signifies that we are quite sure there is no relation. A value close to 0.0 indicates uncertainty, i.e., there may be a relation, but there is no strong evidence for it.

For example, Table 1 shows an example causal activity matrix for the event log L_1. It shows that we are confident that casual relations exists from a to b, from a to c, from b to c, and from c to d, that we are uncertain about a causal relation from b to b, and that we are confident that other causal relations do not exist.

Table 1. Example causal activity matrix M_1 for event log L_1.

From\To	a	b	c	d
a	−0.46	0.88	0.75	−1.00
b	−1.00	0.00	0.88	−1.00
c	−1.00	−1.00	−0.90	1.00
d	−1.00	−1.00	−1.00	−0.67

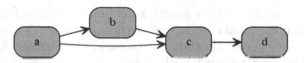

Fig. 1. Example causal activity graph G_1 for causal activity matrix M_1.

Definition 7 (Causal Activity Graph). *Let $A \subseteq \mathcal{U}_A$ be a set of activities. $\mathcal{G}(A)$ denotes the set of causal activity graphs over A. A causal activity graph $G \in \mathcal{G}(A)$ is a 3-tuple $G = (V, E, w)$ where $V \subseteq A$ is the set of nodes, $E \subseteq (V \times V)$ is the set of edges, and $w \in E \to (0.0, 1.0]$ is a weight function that maps every edge onto a positive weight. $G = (V, E, w) \in \mathcal{G}(A)$ is the causal activity graph based on $M \subset \mathcal{M}(A)$ and a specific causality threshold $\tau \in [-1.0, 1.0)$ iff*

- $E = \{(a_1, a_2) \in A \times A \mid M(a_1, a_2) > \tau\}$,
- $V = \bigcup_{(a_1, a_2) \in E} \{a_1, a_2\}$, *and*
- $w((a_1, a_2)) = \frac{M(a_1, a_2) - \tau}{1 - \tau}$ *for $(a_1, a_2) \in E$.*

That is, for every pair of activities $(a_1, a_2) \in A$, there's an edge with a positive weight from a_1 to a_2 in G iff the value for a_1 to a_2 in the causal activity matrix M exceeds some threshold τ. Note that $V \subseteq A$ since some activities in A might not be represented in G.

For example, Fig. 1 shows the causal activity graph that was obtained from the causal activity matrix M_1 using $\tau = 0$.

Definition 8 (Activity Clustering). *Let $A \subseteq \mathcal{U}_A$ be a set of activities. $\mathcal{C}(A)$ denotes the set of activity clusters over A. An activity cluster $C \in \mathcal{C}(A)$ is a subset of A, that is, $C \subseteq A$. $\widehat{\mathcal{C}}(A)$ denotes the set of activity clusterings over A. An activity clustering $\widehat{C} \in \widehat{\mathcal{C}}(A)$ is a set of activity clusters, that is, $\widehat{C} \subseteq \mathcal{P}(A)$[1]. A k-clustering $\widehat{C} \in \widehat{\mathcal{C}}(A)$ is a clustering with size k, i.e. $|\widehat{C}| = k$. The number of activities in $\widehat{C}$ is denoted by $\|\widehat{C}\| = |\bigcup_{C \in \widehat{C}} C|$, i.e. $\|\widehat{C}\|$ signifies the number of unique activities in $\widehat{C}$.*

For example, Fig. 2 shows the activity clustering $\widehat{C}_1$ with size 2 for the causal activity graph G_1. Cluster 0 contains the activities c (as input) and d (as output),

---

[1] $\mathcal{P}(A)$ denotes the powerset over A.

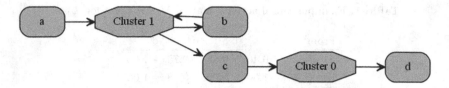

Fig. 2. Example activity clustering $\widehat{C}_1$ for causal activity graph G_1.

whereas Cluster 1 contains a (as input), b (as input and as output), and c (as output). Note that the inputs and output are not part of the definition, but they are included here to better illustrate the fabric of the clusters. Also note that $||\widehat{C}_1|| = |\{a, b, c\} \cup \{c, d\}| = |\{a, b, c, d\}| = 4$.

3.4   Process Models and Process Discovery

Process discovery aims at discovering a model from an event log while conformance checking aims at diagnosing the differences between observed and modeled behavior (resp. the event log and the model). Various discovery algorithms have been proposed in literature. Literature suggests many different notations for models. We abstract from any specific model notation, but will define the set of algorithms that *discover* a model from an event log. These discovery algorithms are often called *mining algorithms*, or *miners* in short.

Definition 9 (Process Model). *Let $A \subseteq \mathcal{U}_A$ be a set of activities. $\mathcal{N}(A)$ denotes the set of process models over A, irrespective of the specific notation (Petri nets, transition systems, BPMN, UML ASDs, etc.) used.*

Definition 10 (Discovery Algorithm). *Let $A \subseteq \mathcal{U}_A$ be a set of activities. $\mathcal{D}(A) = \mathcal{B}(\mathcal{S}(A)) \to \mathcal{N}(A)$ denotes the set of discovery algorithms over A. A discovery algorithm $D \in \mathcal{D}(A)$ discovers a process model over A from an event log over A.*

For example, Fig. 3 shows the Petri net N_1 which was discovered from the event log L_1 using the "ILP-Based Process Discovery" algorithm[2].

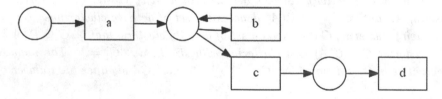

Fig. 3. Example Petri net N_1 discovered from the event log L_1.

---

[2] This algorithm is available in ProM 6.5, see Subsect. 4.3.

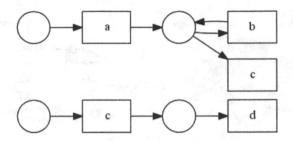

Fig. 4. Submodels discovered for the sublogs of L_1.

3.5 Decomposed Process Discovery

As discussed, in [3], a general approach to decomposed process mining is proposed. In terms of decomposed process discovery, this approach can be explained
as follows: Let $A \subseteq \mathcal{U}_A$ be a set of activities, and let $L \in \mathcal{B}(\mathcal{S}(A))$ be an event log
over A. In order to decompose the activities in L, first a causal activity matrix
$M \in \mathcal{M}(A)$ is *discovered* (cf. Table 1). Any causal activity matrix discovery
algorithm $D_{CA} \in \mathcal{B}(\mathcal{S}(A)) \rightarrow \mathcal{M}(A)$ can be used. From M a causal activity
graph $G \in \mathcal{G}(A)$ is *filtered* (using a specific causality threshold, cf. Fig. 1). By
choosing the value of the causality threshold carefully, we can filter out uncom-
mon causal relations between activities or relations of which we are unsure, for
example those relations introduced by noise in the event log.

Once the causal activity graph G has been constructed, an activity clus-
tering $\widehat{C} \in \widehat{\mathcal{C}}(A)$ is created (cf. Fig. 2). Any activity clustering algorithm
$AC \in \mathcal{G}(A) \rightarrow \widehat{\mathcal{C}}(A)$ can be used to create the clusters. Effectively, an activity clustering algorithm partitions the causal activity graph into partially over-
lapping clusters of activities. Many graph partitioning algorithms have been
proposed in literature [7]. Most algorithms however partition the graph into
non-overlapping subgraphs, while in our case some overlap is required in order
to merge submodels later on in the process. In [3], the so called *maximal decomposition* is used, where the causal activity graph is cut across its vertices and
each edge ends up in precisely one submodel, according to the method proposed
in [15]. This leads to the smallest possible submodels.

Next, for every cluster in the clustering, L is *filtered* to a corresponding
sublog by projecting the cluster to L, i.e., for all $C \in \widehat{C}$ a sublog $L{\restriction}_C$ is created.
For example, based on the activity cluster array $\widehat{C}_1$ (cf. Fig. 2), the event log
L_1 would be split into sublogs $L_1{\restriction}_{\{a,b,c\}}$ and $L_1{\restriction}_{\{c,d\}}$, referring to clusters 1
and 0 respectively. A process model is *discovered* for each sublog. These are
the submodels. Any discovery algorithm $D \in \mathcal{D}(A)$ can be used to discover the
submodels. For example, Fig. 4 shows the submodels that may be discovered
from the sublogs of L_1 as mentioned above.

Finally, the submodels are merged into an overall model (cf. Fig. 3). Any
merging algorithm in $\mathcal{B}(\mathcal{N}(A)) \rightarrow \mathcal{N}(A)$ can be used for this step. Currently,
submodels are merged based on activity labels. Note that we have $|\widehat{C}|$ clusters,

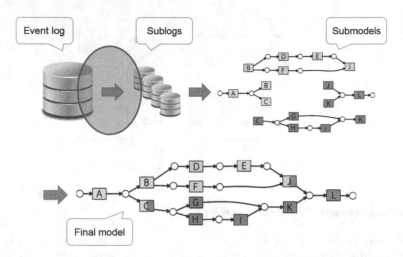

Fig. 5. The general decomposed process discovery workflow, using a more complex event log L_2 as an example. Finding a suitable activity clustering is of key importance.

sublogs and submodels, and $||\widehat{C}||$ activities in the final, merged model. Figure 5 shows the general decomposed process discovery workflow.

This workflow clarifies the generality of the approach. The necessary steps in the approach are defined by their input and output, but any applicable algorithm can be used to perform each step. This also shows the strength of a generic decomposed process mining approach. It might be of interest to only discover a small fragment of a particular process, e.g. only show the part of a medical process where repeated treatment is necessary, or only show the activities (events) performed by some selected resource (provided resource-data is stored in the log). The same holds for decomposed conformance checking. Clustering the activities such that the activities belonging to the interesting part of a process are clustered together will lead to a "filtered" sublog. Applying a fast discovery algorithm to non-interesting parts of the process or only discovering or checking conformance of the submodel of interest can greatly reduce calculation time.

4   A Better Decomposition

It is apparent that the manner in which activities are clustered has a substantial effect on required processing time, and it is possible for similarly sized clusterings (in the average cluster size) to lead to very different total processing times. As a result of the vertex-cut (*maximal*) decomposition approach [3], most activities will be in two (or more) activity clusters, leading to double (or more) work, as the clusters have a lot of overlap and causal relations between them, which might not be desirable. From the analysis results in [13] we can see that this introduces a lot of unwanted overhead, and generally reduces model quality. Also, sequences or sets of activities with high causal relations are generally easily (and thus quickly) discovered by process discovery algorithms, yet the approach will often split up these activities over different clusters.

Model quality can potentially suffer from a decomposition that is too fine-grained. It might be that the sublogs created by the approach contain too little information for the process discovery algorithm to discover a good, high quality submodel from, or that a process is split up where it shouldn't be. Merging these low-quality submodels introduces additional problems.

In order to achieve a high quality process model in little time, we have to find a decomposition that produces high quality submodels with as little overlap as possible (shared activities) and causal relations between them. Also, the submodels should preferably be of comparable size, because of the exponential nature of most process discovery algorithms [2]. Hence, a *good* decomposition should (1) *maximize* the causal relations between the activities *within* each cluster in the activity clustering, (2) *minimize* the causal relations and overlap *across* the clusters and (3) have approximately *equally sized* clusters. The challenge lies in finding a good balance between these three clustering properties. In Subsect. 4.1, we formally define these properties and provide metrics in order to be able to asses the quality of a given decomposition before using it to discover a model or check conformance with.

A clustering where one cluster is a subset of another cluster is not *valid* as it would lead to double work, and would thus result in an increase in required processing time without increasing (or even decreasing) model quality. Note that this definition of a valid clustering allows for disconnected clusters, and that some activities might not be in any cluster. This is acceptable as processes might consist of disconnected parts and event logs may contain noise. However, if activities are left out some special processing might be required.

Definition 11 (Valid Clustering). *Let $A \subseteq \mathcal{U}_A$ be a set of activities. Let $\widehat{C} \in \widehat{\mathcal{C}}(A)$ be a clustering over A. $\widehat{C}$ is a valid clustering iff: $\widehat{C} \neq \emptyset \wedge \forall_{C_1,C_2 \in \widehat{C} \wedge C_1 \neq C_2} C_1 \not\subseteq C_2$. $\widehat{\mathcal{C}}_\mathcal{V}(A)$ denotes the set of valid clusterings over A.*

For example, the clustering shown in Fig. 2 is valid, as $\{a, b, c\} \not\subseteq \{c, d\}$ and $\{c, d\} \not\subseteq \{a, b, c\}$.

4.1   Clustering Properties

We define decomposition quality notions in terms of clustering properties.

The first clustering property we define is *cohesion*. The cohesion of an activity clustering is defined as the average cohesion of each activity cluster in that clustering. A clustering with good cohesion (cohesion ≈ 1) signifies that causal relations between activities in the same cluster are optimized, whereas bad cohesion (cohesion ≈ 0) signifies that activities with few causal relations are clustered together.

Definition 12 (Cohesion). *Let $A \subseteq \mathcal{U}_A$ be a set of activities. Let $G = (V, E, w) \in \mathcal{G}(A)$ be a causal activity graph over A, and let $\widehat{C} \in \widehat{\mathcal{C}}_\mathcal{V}(A)$ be* a valid clustering over A. The cohesion of clustering $\widehat{C}$ in graph G, denoted Cohesion($\widehat{C}, G$), is defined as follows:

$$Cohesion(\widehat{C}, G) = \frac{\sum_{C \in \widehat{C}} Cohesion(C, G)}{|\widehat{C}|}$$

$$Cohesion(C, G) = \frac{\sum_{(a_1, a_2) \in E \cap (C \times C)} w((a_1, a_2))}{|C \times C|}$$

For example:

- $Cohesion(\{a, b, c\}, G_1) = (0.88 + 0.75 + 0.88)/9 = 0.28$,
- $Cohesion(\{c, d\}, G_1) = 1.00/4 = 0.25$, and
- $Cohesion(\widehat{C}_1, G_1) = (0.28 + 0.25)/2 = 0.26$.

The second clustering property is called *coupling*, and is also represented by a number between 0 and 1. Good coupling (coupling ≈ 1) signifies that causal relations between activities across clusters are minimized. Bad coupling (coupling ≈ 0) signifies that there are a lot of causal relations between activities in different clusters.

Definition 13 (Coupling). Let $A \subseteq \mathcal{U}_A$ be a set of activities. Let $G = (V, E, w) \in \mathcal{G}(A)$ be a causal activity graph over A, and let $\widehat{C} \in \widehat{C}_V(A)$ be a valid clustering over A. The coupling of clustering $\widehat{C}$ in graph G, denoted Coupling($\widehat{C}, G$), is defined as follows:

$$Coupling(\widehat{C}, G) = \begin{cases} 1 & \text{if } |\widehat{C}| \leq 1 \\ 1 - \dfrac{\sum\limits_{C_1, C_2 \in \widehat{C} \wedge C_1 \neq C_2} Coupling(C_1, C_2, G)}{|\widehat{C}| \cdot (|\widehat{C}| - 1)} & \text{if } |\widehat{C}| > 1 \end{cases}$$

$$Coupling(C_1, C_2, G) = \frac{\sum\limits_{(a_1, a_2) \in E \cap ((C_1 \times C_2) \cup (C_2 \times C_1))} w((a_1, a_2))}{2 \cdot |C_1 \times C_2|}$$

For example:

- $Coupling(\{a, b, c\}, \{c, d\}, G_1) = (0.75 + 0.88 + 1.00)/12 = 0.22$,
- $Coupling(\{c, d\}, \{a, b, c\}, G_1) = (0.75 + 0.88 + 1.00)/12 = 0.22$, and
- $Coupling(\widehat{C}_1, G_1) = 1 - (0.22 + 0.22)/2 = 0.78$.

Note that the weights of the causal relations are used in the calculation of cohesion and coupling. Relations of which we are not completely sure of (or that are weak) therefore have less effect on these properties than stronger ones.

The *balance* of an activity clustering is the third property. A clustering with good balance has clusters of (about) the same size. Like cohesion and coupling, balance is also represented by a number between 0 and 1, where a good balance (balance ≈ 1) signifies that all clusters are about the same size and a bad balance (balance ≈ 0) signifies that the cluster sizes differ quite a lot. Decomposing the activities into clusters with low balance (e.g. a k-clustering with one big cluster holding almost all of the activities and $(k-1)$ clusters with only a few activities) will not speed up discovery or conformance checking, rendering the whole decomposition approach useless. At the same time finding a clustering with perfect balance (all clusters have the same size) will most likely split up the process/log in places that "shouldn't be split up", as processes generally consist out of different-sized natural parts.

This balance formula utilizes the standard deviation of the sizes of the clusters in a clustering to include the magnitude of the differences in cluster sizes. A variation of this formula using squared errors or deviations could also be used as a clustering balance measure.

Definition 14 (Balance). *Let $A \subseteq \mathcal{U}_A$ be a set of activities. Let $\widehat{C} \in \widehat{\mathcal{C}}_{\mathcal{V}}(A)$ be a valid clustering over A. The balance of clustering $\widehat{C}$ denoted $Balance(\widehat{C})$ is defined as follows:*

$$Balance(\widehat{C}) = 1 - \frac{2 \cdot \sigma(\widehat{C})}{||\widehat{C}||}$$

Where $\sigma(\widehat{C})$ signifies the standard deviation of the sizes of the clusters in the clustering $\widehat{C}$.

For example, $Balance(\widehat{C}_1) = 1 - (2 \cdot 0.5)/2 = 0.5$.

In order to assess a certain decomposition based on the clustering properties, we use a weighted scoring function, which grades an activity clustering with a score between 0 (bad clustering) and 1 (good clustering). A weight can be set for each clustering property, depending on their relative importance. A clustering with clustering score 1 therefore has perfect cohesion, coupling and balance scores, on the set weighing of properties.

Definition 15 (Clustering Score). *Let $A \subseteq \mathcal{U}_A$ be a set of activities. Let $G \in \mathcal{G}(A)$ be a causal activity graph over A, and let $\widehat{C} \in \widehat{\mathcal{C}}_{\mathcal{V}}(A)$ be a valid clustering over A. The clustering score (score) of clustering $\widehat{C}$ in graph G, denoted $Score(\widehat{C}, G)$, is defined as follows:*

$$Score(\widehat{C}, G) = Cohesion(\widehat{C}, G) \cdot \left(\frac{Coh_W}{Coh_W + Cou_W + Bal_W} \right)$$
$$+ Coupling(\widehat{C}, G) \cdot \left(\frac{Cou_W}{Coh_W + Cou_W + Bal_W} \right)$$
$$+ Balance(\widehat{C}) \cdot \left(\frac{Bal_W}{Coh_W + Cou_W + Bal_W} \right)$$

where Coh_W, Cou_W, and Bal_W are the weights for Cohesion, Coupling, and Balance.

For example, if we take all weights to be 10, that is, $Coh_W = Cou_W = Bal_W = 10$, then $Score(\widehat{C}_1, G_1) = 0.26 \cdot (10/30) + 0.78 \cdot (10/30) + 0.5 \cdot (10/30) = 0.51$.

4.2 Recomposition of Activity Clusters

As described in Subsect. 3.5, creating a good activity clustering is essentially a graph partitioning problem. The causal activity graph needs to be partitioned in parts that have (1) good cohesion, (2) good coupling and (3) good balance. The existing *maximal* decomposition approach [3] often leads to a decomposition that is too decomposed, i.e. too fine-grained. Cohesion and balance of clusterings found by this approach are usually quite good, since all clusters consist of only a few related activities. However, coupling is inherently bad, since there's a lot of overlap in the activity clusters and there are many causal relations across clusters. This decomposition approach leads to unnecessary and unwanted overhead and potential decreased model quality. We thus want to find a possibly *non-maximal* decomposition which optimizes the three clustering properties.

Instead of applying or creating a different graph partitioning algorithm, we *recompose* the activity clusters obtained by the vertex-cut decomposition, since it is maximal (no smaller valid clustering exists [3]). The idea is that it is possible to create a clustering that has fewer, larger clusters, requiring less processing time to discover the final model, because overhead as well as cluster overlap are reduced. Additionally, model quality is likely to increase because of the higher number of activities in each cluster and the lower coupling between clusters.

For example, if we put all activities from event log L_1 in a single cluster $\{a, b, c, d\}$, yielding clustering $\widehat{C}'_1$, and use the same weights, then we would get the following scores:

- $Cohesion(\widehat{C}'_1, G_1) = (0.88 + 0.75 + 0.88 + 1.00)/16 = 0.22$,
- $Coupling(\widehat{C}'_1, G_1) = 1$,
- $Balance(\widehat{C}'_1) = 1$, and
- $Score(\widehat{C}'_1, G_1) = 0.22 \cdot (10/30) + 1 \cdot (10/30) + 1 \cdot (10/30) = 0.74$.

Clearly, as $0.74 > 0.51$, the chosen weights would lead to the situation where a single cluster $\{a, b, c, d\}$ would be preferred over two clusters $\{a, b, c\}$ and $\{c, d\}$.

There are often many ways in which a clustering can be recomposed to the desired amount of clusters, as shown in Fig. 6. We are interested in the highest quality clustering of the desired size k, i.e. the k-clustering that has the best cohesion, coupling and balance properties. A k-clustering that has a high clustering score will very likely lead to such a decomposition.

In order to find a good decomposition in the form of a high-scoring clustering quickly, we propose two agglomerative hierarchical recomposition approaches, which iteratively merge clusters, reducing the size of the clustering by one each iteration. As the amount of k-clusterings for a given causal activity graph is finite, it is possible to exhaustively find the best k-clustering. However, for even moderately-sized event logs (in the number of activities) this is too resource- and time-consuming, as shown in [13]. Also, a semi-exhaustive "random" recomposition approach was implemented that randomly recomposes the clustering to k clusters a given amount of times and returns the highest-scoring k-clustering found. This method is used as a benchmark for the hierarchical recomposition approaches.

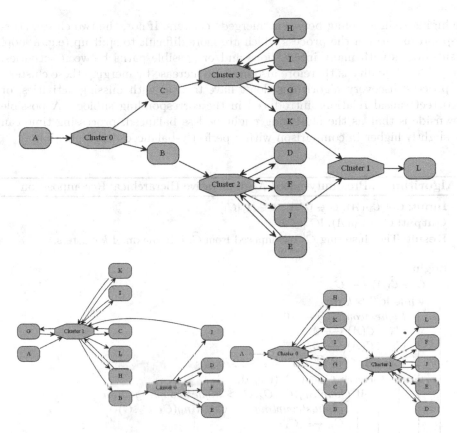

Fig. 6. 2 possible recompositions for event log L_2 (cf. Fig. 5) from 4 (top) to 2 clusters (bottom left and right). Finding a good recomposition is key to creating a coarser clustering which could potentially decrease processing time and increase model quality.

Proximity-Based Approach. We propose an hierarchical recomposition approach based on proximity between activity clusters, where cluster coupling is used as the proximity measure (Algorithm 1). The starting point is the clustering as created by the vertex-cut approach. We repeatedly merge the clusters closest to one another (i.e. the pair of clusters with the highest coupling) until we end up with the desired amount of clusters (k). After the k-clustering is found, it is made valid by removing any clusters that are a subcluster of another cluster, if such clusters exist. It is therefore possible that the algorithm returns a clustering with size smaller than k.

By merging clusters we are likely to lower the overall cohesion of the clustering. This drawback is minimized, as coupling is used as the distance measure. Coupling is also minimized. The proximity-based hierarchical recomposition approach however is less favored towards the balance property, as it is possible that -because of high coupling between clusters- two of the larger clusters are merged. In most processes however, coupling between two "original" clusters will be higher than coupling between "merged" clusters. If not, the two clusters correspond to parts of the process which are more difficult to split up (e.g. a loop, a subprocess with many interactions and/or possible paths between activities, etc.). Model quality is therefore also likely to increase by merging these clusters, as process discovery algorithms don't have to deal with missing activities, or incorrect causal relations introduced in the corresponding sublogs. A possible downside is that as the clustering might be less balanced, processing time can be slightly higher in comparison with a perfectly-balanced decomposition.

Algorithm 1. Proximity-based Agglomerative Hierarchical Recomposition

Input: $\widehat{C} \in \widehat{\mathcal{C}}_V(A)$, $G \in \mathcal{G}(A)$, $k \in [1, |\widehat{C}|]$
Output: $\widehat{C}' \in \widehat{\mathcal{C}}_V(A)$, $|\widehat{C}'| \leq k$
Result: The clustering $\widehat{C}'$ recomposed from $\widehat{C}$, into maximal k clusters.

begin
  $\widehat{C}' \in \widehat{\mathcal{C}}(A) \longleftarrow \widehat{C}$
  while $|\widehat{C}'| > k$ **do**
    $highestcoupling \longleftarrow 0$
    $C_A \in \mathcal{C}(A) \longleftarrow \emptyset$
    $C_B \in \mathcal{C}(A) \longleftarrow \emptyset$
    foreach $C_1 \in \widehat{C}'$ **do**
      foreach $C_2 \in \widehat{C}' \setminus \{C_1\}$ **do**
        if $Coupling(C_1, C_2, G) > highestcoupling$ **then**
          $highestcoupling \longleftarrow Coupling(C_1, C_2, G)$
          $C_A \longleftarrow C_1$
          $C_B \longleftarrow C_2$

    $\widehat{C}' \longleftarrow \widehat{C}' \setminus \{C_A, C_B\} \bigcup \{C_A \cup C_B\}$
    $\widehat{C}'' \in \widehat{\mathcal{C}}(A) \longleftarrow \widehat{C}'$
    foreach $C \in \widehat{C}'' \setminus \{C_A \cup C_B\}$ **do**
      if $C \subseteq \{C_A \cup C_B\}$ **then**
        $\widehat{C}' \longleftarrow \widehat{C}' \setminus \{C\}$

  return $\widehat{C}'$

For example, Fig. 7 shows the 2-clustering that is obtained by this approach when we start from the 4-clustering for event log L_2 shown earlier. Cluster 0 is merged with Cluster 3, and Cluster 1 with Cluster 2.

Score-Based Approach. We propose a second hierarchical recomposition algorithm that uses the scoring function in a look-ahead fashion (Algorithm 2). In essence, this algorithm, like the proximity-based variant, iteratively merges two clusters into one. For each combination of clusters, the score of the clustering that

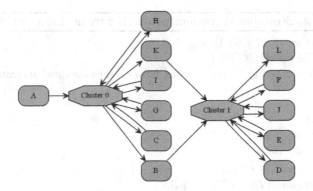

Fig. 7. Best 2-clustering found when starting from the 4-clustering shown at the top of Fig. 6 using the Proximity-based approach.

results from merging those clusters is calculated. The clustering with the highest score is used for the next step. The algorithm is finished when a k-clustering is reached. Like in the proximity-based approach, after the k-clustering is found, it is made valid by removing any clusters that are a subcluster of another cluster, if such clusters exist.

The advantage of this approach is that specific (combinations of) clustering properties can be given priority, by setting their scoring weight(s) accordingly. For example, it is possible to distribute the activities over the clusters near perfectly, by choosing a high relative weight for balance. This would likely lead to a lower overall processing time. However, it might lead to natural parts of the process being split over multiple clusters, which could negatively affect model quality. A downside of this algorithm is that, as the algorithm only looks ahead one step, it is possible that a choice is made that ultimately leads to a lower clustering score, as that choice cannot be undone in following steps.

For example, Fig. 8 shows the 2-clustering that is obtained by this approach when we start from the 4-clustering for event log L_2 shown earlier where all weights have been set to 10. Cluster 0 is now merged with Cluster 1, and Cluster 2 with Cluster 3.

4.3 Implementation

All concepts and algorithms introduced in this paper are implemented in release 6.5 of the process mining toolkit *ProM*[3] [24], developed at the Eindhoven University of Technology. All work can be found in the *ActivityClusterArrayCreator* package, which is part of the *DivideAndConquer* package suite. This suite is installed by default in ProM 6.5.

The *plug-in* to use in ProM is *Modify Clusters*, which takes an activity cluster array (clustering) and a causal activity graph as input. The user can decide to

---

[3] ProM 6.5 can be downloaded from http://www.promtools.org.

Algorithm 2. Score-based Agglomerative Hierarchical Recomposition

Input: $\widehat{C} \in \widehat{\mathcal{C}}_{\mathcal{V}}(A)$, $G \in \mathcal{G}(A)$, $k \in [1, |\widehat{C}|]$
Output: $\widehat{C}' \in \widehat{\mathcal{C}}_{\mathcal{V}}(A)$, $|\widehat{C}'| \leq k$
Result: The clustering $\widehat{C}'$ recomposed from $\widehat{C}$, into maximal k clusters.

begin
  $\widehat{C}' \in \widehat{\mathcal{C}}(A) \longleftarrow \widehat{C}$
  while $|\widehat{C}'| > k$ **do**
    $highestscore \longleftarrow 0$
    foreach $C_1 \in \widehat{C}'$ **do**
      foreach $C_2 \in \widehat{C}' \setminus \{C_1\}$ **do**
        $\widehat{C}'' \longleftarrow \widehat{C}' \setminus \{C_1, C_2\} \bigcup \{C_1 \cup C_2\}$
        if $Score(\widehat{C}'', G) > highestscore$ **then**
          $highestscore \longleftarrow Score(\widehat{C}'', G)$
          $\widehat{C}''' \in \widehat{\mathcal{C}}(A) \longleftarrow \widehat{C}''$
    $\widehat{C}' \longleftarrow \widehat{C}'''$
  $\widehat{C}'' \longleftarrow \widehat{C}'$
  foreach $C_1 \in \widehat{C}''$ **do**
    foreach $C_2 \in \widehat{C}'' \setminus \{C_1\}$ **do**
      if $C_1 \subseteq C_2$ **then**
        $\widehat{C}' \longleftarrow \widehat{C}' \setminus \{C_1\}$
  return $\widehat{C}'$

either set the parameters of the action using a dialog, or to use the default parameter values. Figure 9 shows both options, where the top one will use the dialog and the bottom one the default parameter values. Figure 10 shows the dialog. At the top, the user can select which approach to use:

Brute Force: Use a brute force approach to find the optimal k-clustering.
Incremental using Best Coupling: Use the proximity-based approach to find a k-clustering.
Incremental using Best Coupling (only Overlapping Clusters): Use the proximity-based approach to find a k-clustering, but allow only to merge clusters that actually overlap (have an activity in common). This is the default approach.
Incremental using Best Score: Use the score-based approach to find a k-clustering.
Incremental using Best Score (only Overlapping Clusters): Use the score-based approach to find a k-clustering, but allow only to merge clusters that actually overlap (have an activity in common).
Random: Randomly merge clusters until a k-clustering is reached.

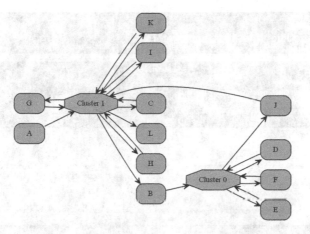

Fig. 8. Best 2-clustering found when starting from the 4-clustering shown at the top of Fig. 6 using the Score-based approach with all weights set to 10.

Fig. 9. The *Modify Clusters* actions in ProM 6.5.

Below, the user can set the value of k, which is set to 50 % of the number of existing clusters by default. At the bottom, the user can set the respective weights to a value from 0 to 100. By default, all weights are set to 100.

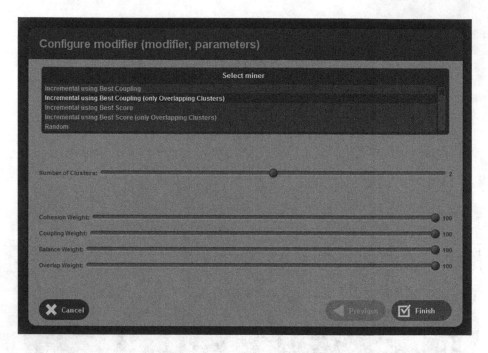

Fig. 10. The dialog that allows the user to set parameter values.

5   Use Case

The proposed recomposition techniques are tested using event logs of different sizes and properties. Results for an event log consisting of 33 unique activities, and 1000 traces are shown in this section. For this test the ILP Miner process discovery algorithm was used [25]. Discovering a model directly for this log will lead to a high quality model, but takes ~25 min on a modern quad-core system [13]. The vertex-cut decomposed process mining approach is able to discover a model in roughly 90 s, however the resulting model suffers from disconnected activities (i.e. a partitioned model). The goal is thus to find a balance between processing times and model quality.

We are interested in the clustering scores of each algorithm when recomposing the clustering created by the vertex-cut approach to a smaller size. Exhaustively finding the best possible clustering proved to be too time- and resource-consuming. Therefore, besides the two approaches listed here, a random recomposition approach was used which recomposes clusters randomly one million times. The highest found score is shown as to give an idea of what the best possible clustering might be. Equal weights were used for the three clustering properties in order to compute the clustering scores. All clustering scores are shown in Fig. 11. As can be seen, the vertex-cut approach creates 22 clusters. We can see that all algorithms perform very similarly in terms of clustering score.

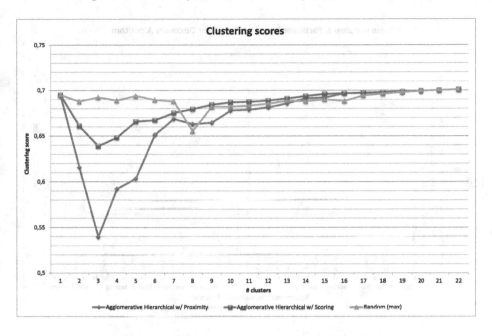

Fig. 11. Clustering score per recomposition algorithm.

Only for very small clustering sizes the proximity-based approach performs worse than the other approaches, due to its tendency to create unbalanced clusters.

Besides clustering scores, we are even more interested in how each decomposition method performs in terms of required processing time and quality of the resulting process model. In Fig. 12 we can see that decomposing the event log drastically reduces processing times. For an event log this size, the decomposition steps relatively takes up negligible time (see base of bars in figure), as most time is spent discovering the submodels (light blue bars). Processing times are reduced exponentially (as expected), until a certain optimum decomposition (in terms of speed) is reached, after which overhead starts to increase time linearly again.

We have included two process models (Petri Nets) discovered from the event log. Figure 14 shows the model discovered when using the vertex-cut decomposition. Figure 15 shows the model discovered when using the clustering recomposed to 11 clusters with the Proximity-based agglomerative hierarchical approach. We can see that in Fig. 14, activity "10" is disconnected (marked blue). In Fig. 15, this activity is connected, and a structure (self-loop) is discovered. We can also see that activities "9" and "12" now are connected to more activities. This shows that the vertex-cut decomposition sometimes splits up related activities, which leads to a lower quality model. Indeed, Fig. 13 shows that the activities "9", "10", and "12" were split over two clusters in the vertex-cut decomposition (top), and were regrouped (bottom) by our recomposition. By recomposing the clusters we

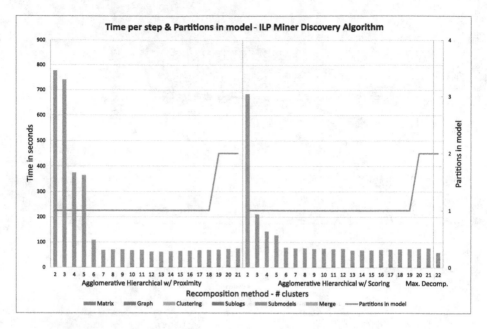

Fig. 12. Time per step & Partitions in model using the agglomerative hierarchical recomposition approaches and the ILP Miner process discovery algorithm (Color figure online).

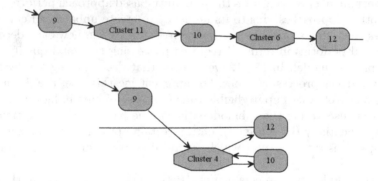

Fig. 13. Activities that were split over multiple clusters by the vertex-cut decomposition (top), that were regrouped by our recomposition (bottom).

rediscover these relations, leading to a higher quality model. Processing times for these two models are comparable, as can be seen in Fig. 12.

The proposed agglomerative hierarchical recomposition algorithms are able to create activity clusterings that have good *cohesion, coupling* and *balance* properties in very little time. Often times the scores of these clusterings are almost as good as the scores of clusterings created by the much slower exhaustive approaches [13]. The results show that by creating such good clusterings,

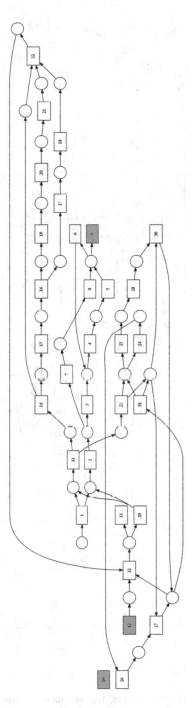

Fig. 14. Process model discovered using the vertex-cut decomposition. Some activities are disconnected in the final model (Color figure online).

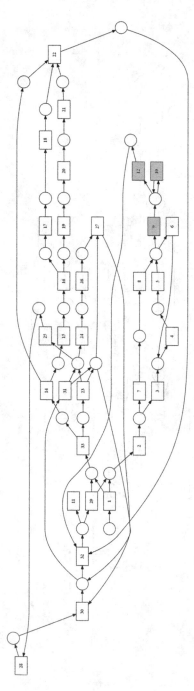

Fig. 15. Process model discovered using the vertex-cut clustering recomposed to 11 clusters. Previously disconnected activities are connected again, improving model quality (Color figure online).

indeed a better decomposition can be made. Because cluster overlap and the amount of clusters are reduced, overhead can be minimized, reducing required processing time. This results in a comparable total required processing time from event log to process model, even though an extra step is necessary. The larger the event log, the higher the time gain. By recomposing the clusters, processes are split up less in places where they shouldn't be, leading to better model quality. Because the discovered submodels are of higher quality and have less overlap and coupling, the merging step introduces less or no unnecessary (implicit) or double paths in a model, which leads to improvements in precision, generalization and simplicity of the final model.

6   Conclusions and Future Work

In decomposed process discovery, large event logs are decomposed by somehow clustering their events (activities), and there are many ways these activity clusterings can be made. Hence, good quality notions are necessary to be able to assess the quality of a decomposition before starting the time-consuming actual discovery algorithm. Being able to find a high-quality decomposition plays a key role in the success of decomposed process mining, even though the decomposition step takes relatively very little time.

By using a better decomposition, less problems arise when discovering submodels for sublogs and when merging submodels into the overal process model. We introduced three quality notions in the form of clustering properties: *cohesion*, *coupling* and *balance*. It was shown that finding a *non-maximal* decomposition can potentially lead to a decrease in required processing time while maintaining or even improving model quality, compared to the existing vertexcut *maximal* decomposition approach. We have proposed two variants of an agglomerative hierarchical recomposition technique, which are able to create a high-quality decomposition for any given size, in very little time.

Even though the scope was limited to decomposed process discovery, the introduced quality notions and decomposition approaches can be applied to decomposed conformance checking as well. However, more work is needed to incorporate them in a conformance checking environment.

Besides finding a better decomposition, we believe improvements can be gained in finding a better, more elaborate algorithm to merge submodels into the overal process model. By simply merging submodels based on activity labels it is likely that implicit paths are introduced. Model quality in terms of fitness, simplicity, generality or precision could suffer. An additional post-processing step (potentially using causal relations) could also solve this issue.

Even though most interesting process discovery algorithms are exponential in the number of different activities, adding an infrequent or almost unrelated activity to a cluster might not increase computation time for that cluster as much as adding a frequent or highly related one. Therefore, besides weighing causal relations between activities in the causal activity matrix, activities themselves might be weighted as well. Frequency and connectedness are some of the many possible properties that can be used as weights. It might be possible that one part of a process can be discovered easily by a simple algorithm whereas another, more complex part of the process needs a more involved discovery algorithm to be modeled correctly. Further improvements in terms of processing time can be gained by somehow detecting the complexity of a single submodel in a sublog, and choosing an adequate discovery algorithm.

Finally, as discussed, the proposed recomposition algorithms expect the desired amount of clusters to be given. Even though the algorithms were shown to provide good results for any chosen number, the approach would benefit from some method that determines a fitting clustering size for a given event log. This would also mean one less potentially uncertain step for the end-user.

References

1. van der Aalst, W.M.P.: Process Mining: Discovery, Conformance and Enhancement of Business Processes. Springer, Berlin (2011)
2. van der Aalst, W.M.P.: Distributed process discovery and conformance checking. In: de Lara, J., Zisman, A. (eds.) FASE 2012. LNCS, vol. 7212, pp. 1–25. Springer, Heidelberg (2012)
3. van der Aalst, W.M.P.: A general divide and conquer approach for process mining. In: 2013 Federated Conference on Computer Science and Information Systems (FedCSIS), pp. 1–10. IEEE (2013)
4. van der Aalst, W.M.P.: Decomposing Petri nets for process mining: a generic approach. Distrib. Parallel Databases **31**(4), 471–507 (2013)
5. van der Aalst, W.M.P., et al.: Process mining manifesto. In: Daniel, F., Barkaoui, K., Dustdar, S. (eds.) BPM Workshops 2011, Part I. LNBIP, vol. 99, pp. 169–194. Springer, Heidelberg (2012)
6. van der Aalst, W.M.P., Verbeek, H.M.W.: Process discovery and conformance checking using passages. Fundamenta Informaticae **131**(1), 103–138 (2014)
7. Buluç, A., Meyerhenke, H., Safro, I., Sanders, P., Schulz, C.: Recent advances in graph partitioning. CoRR abs/1311.3144 (2013). http://arxiv.org/abs/1311.3144
8. Cannataro, M., Congiusta, A., Pugliese, A., Talia, D., Trunfio, P.: Distributed data mining on grids: services, tools, and applications. IEEE Trans. Syst. Man Cybern. Part B Cybern. **34**(6), 2451–2465 (2004)
9. Carmona, J.: Projection approaches to process mining using region-based techniques. Data Min. Knowl. Discov. **24**(1), 218–246 (2012). http://dblp.uni-trier.de/db/journals/datamine/datamine24.html
10. Carmona, J.A., Cortadella, J., Kishinevsky, M.: A region-based algorithm for discovering petri nets from event logs. In: Dumas, M., Reichert, M., Shan, M.-C. (eds.) BPM 2008. LNCS, vol. 5240, pp. 358–373. Springer, Heidelberg (2008)
11. Carmona, J., Cortadella, J., Kishinevsky, M.: Divide-and-conquer strategies for process mining. In: Dayal, U., Eder, J., Koehler, J., Reijers, H.A. (eds.) BPM 2009. LNCS, vol. 5701, pp. 327–343. Springer, Heidelberg (2009)
12. Goedertier, S., Martens, D., Vanthienen, J., Baesens, B.: Robust process discovery with artificial negative events. J. Mach. Learn. Res. **10**, 1305–1340 (2009)
13. Hompes, B.F.A.: On decomposed process mining: how to solve a Jigsaw puzzle with friends. Master's thesis, Eindhoven University of Technology, Eindhoven, The Netherlands (2014). http://repository.tue.nl/776743

14. Kambatla, K., Kollias, G., Kumar, V., Grama, A.: Trends in big data analytics. J. Parallel Distrib. Comput. **74**(7), 2561–2573 (2014)
15. Kim, M., Candan, K.: SBV-Cut: vertex-cut based graph partitioning using structural balance vertices. Data Knowl. Eng. **72**, 285–303 (2012)
16. Munoz-Gama, J., Carmona, J., van der Aalst, W.M.P.: Single-entry single-exit decomposed conformance checking. Inf. Syst. **46**, 102–122 (2014). http://dx.doi.org/10.1016/j.is.2014.04.003
17. Park, B.H., Kargupta, H.: Distributed data mining: algorithms, systems, and applications, pp. 341–358 (2002)
18. Reguieg, H., Toumani, F., Motahari-Nezhad, H.R., Benatallah, B.: Using mapreduce to scale events correlation discovery for business processes mining. In: Barros, A., Gal, A., Kindler, E. (eds.) BPM 2012. LNCS, vol. 7481, pp. 279–284. Springer, Heidelberg (2012)
19. Saha, B., Shah, H., Seth, S., Vijayaraghavan, G., Murthy, A., Curino, C.: Apache tez: A unifying framework for modeling and building data processing applications. In: Proceedings of the 2015 ACM SIGMOD International Conference on Management of Data, pp. 1357–1369. ACM (2015)
20. Shukla, R.K., Pandey, P., Kumar, V.: Big data frameworks: at a glance. Int. J. Innov. Adv. Comput. Sci. (IJIACS) **4**(1) (2015). ISSN 2347-8616
21. The Apache Software Foundation: Apache Flink: Scalable Batch and Stream Data Processing. http://flink.apache.org/, July 2015
22. Vanderfeesten, I.T.P.: Product-based design and support of workflow processes. Ph.D. thesis, Eindhoven University of Technology, Eindhoven, The Netherlands (2009)
23. Vavilapalli, V.K., Murthy, A.C., Douglas, C., Agarwal, S., Konar, M., Evans, R., Graves, T., Lowe, J., Shah, H., Seth, S., et al.: Apache hadoop yarn: yet another resource negotiator. In: Proceedings of the 4th Annual Symposium on Cloud Computing, p. 5. ACM (2013)
24. Verbeek, H.M.W., Buijs, J.C.A.M., van Dongen, B.F., van der Aalst, W.M.P.: ProM 6: the process mining toolkit. In: Proceedings of BPM Demonstration Track 2010, vol. 615, pp. 34–39. CEUR-WS.org (2010). http://ceur-ws.org/Vol-615/paper13.pdf
25. van der Werf, J.M.E.M., van Dongen, B.F., Hurkens, C.A.J., Serebrenik, A.: Process discovery using integer linear programming. In: van Hee, K.M., Valk, R. (eds.) PETRI NETS 2008. LNCS, vol. 5062, pp. 368–387. Springer, Heidelberg (2008)
26. Zaharia, M., Chowdhury, M., Franklin, M.J., Shenker, S., Stoica, I.: Spark: cluster computing with working sets. In: Proceedings of the 2nd USENIX conference on Hot topics in cloud computing, vol. 10, p. 10 (2010)

History-Based Construction of Alignments for Conformance Checking: Formalization and Implementation

Mahdi Alizadeh$^{(\boxtimes)}$, Massimiliano de Leoni, and Nicola Zannone

Department of Mathematics and Computer Science,
Eindhoven University of Technology, P.O. Box 513,
5600 MB Eindhoven, The Netherlands
{m.alizadeh,m.d.leoni,n.zannone}@tue.nl

Abstract. Alignments provide a robust approach for conformance checking, which has been largely applied in various contexts such as auditing and performance analysis. Alignment-based conformance checking techniques pinpoint the deviations causing nonconformity based on a cost function. However, such a cost function is often manually defined on the basis of human judgment and thus error-prone, leading to alignments that do not provide accurate explanations of nonconformity. This paper proposes an approach to automatically define the cost function based on information extracted from the past process executions. The cost function only relies on objective factors and thus enables the construction of probable alignments, i.e. alignments that provide probable explanations of nonconformity. Our approach has been implemented in ProM and evaluated using both synthetic and real-life data.

Keywords: Conformance checking · Alignments · Cost functions

1 Introduction

Modern organizations are centered on the processes needed to deliver products and services in an efficient and effective manner. Organizations that operate at a higher process maturity level use formal/semiformal models (e.g., UML, EPC, BPMN and YAWL models) to document their processes. In some case these models are used to configure process-aware information systems (e.g., WFM or BPM systems). However, in most organizations process models are not used to enforce a particular way of working. Instead, process models are used for discussion, performance analysis (e.g., simulation), certification, process improvement, etc. However, reality may deviate from such models. People tend to focus on idealized process models that have little to do with reality. This illustrates the importance of *conformance checking* [1,2,12].

This work is an extended and revised version of [8].

© IFIP International Federation for Information Processing 2015
P. Ceravolo et al. (Eds.): SIMPDA 2014, LNBIP 237, pp. 58–78, 2015.
DOI: 10.1007/978-3-319-27243-6_3

Conformance checking aims to verify whether the observed behavior recorded in an event log matches the intended behavior represented as a process model. The notion of alignments [2] provides a robust approach to conformance checking, which makes it possible to pinpoint the deviations causing nonconformity. An alignment between a recorded process execution and a process model is a pairwise matching between activities recorded in the log and activities allowed by the model. Sometimes, activities as recorded in the event log (events) cannot be matched to any of the activities allowed by the model (process activities). For instance, an activity is executed when not allowed. In this case, we match the event with a special *null* activity (hereafter, denoted as ≫), thus resulting in a so-called *move on log*. Other times, an activity should have been executed but is not observed in the event log. This results in a process activity that is matched to a ≫ event, thus resulting in a so-called *move on model*.

Alignments are powerful artifacts to detect nonconformity between the observed behavior as recorded in the event log and the prescribed behavior as represented by process models. In fact, when an alignment between a log trace and process model contains at least one move on log or model, it means that such a log trace does not conform to the model. As a matter of fact, moves on log/model indicate where the execution is not conforming by pinpointing the deviations that have caused this nonconformity.

In general, a large number of possible alignments exist between a process model and a log trace, since there may exist manifold explanations why a trace is not conforming. It is clear that one is interested in finding what really happened. Adriansyah et al. [4] have proposed an approach based on the principle of the Occam's razor: the simplest and most parsimonious explanation is preferable. Therefore, one should not aim to find any alignment but, precisely, one of the alignments with the least expensive deviations (one of the so-called *optimal alignments*), according to some function assigning costs to deviations.

Existing alignment-based conformance checking techniques (e.g. [2,4]) require process analysts to manually define a cost function based on their background knowledge and beliefs. The definition of such a cost function is fully based on human judgment and, thus, prone to imperfections. These imperfections ultimately lead to alignments that are optimal, according to the provided cost function, but that do not provide an explanation of what really happened.

In this paper, we propose an alternative way to define a cost function, where the human judgment is put aside and only objective factors are considered. The cost function is automatically constructed by looking at the logging data and, more specifically, at the past process executions that are compliant with the process model. The intuition behind is that one should look at the past history of process executions and learn from it what are the probable explanations of nonconformity. In particular, probable explanations of nonconformity for a certain process execution can be obtained by analyzing the behavior observed for such a process execution in each and every state and the behavior observed for other confirming traces when they were in the same state. Our approach gives a potentially different cost for each move on model and log (depending on the current state), leading to the definition of a more sensitive cost function.

The approach has been fully implemented as a software plug-in for the open-source process-mining framework *ProM*. To assess the practical relevance of our approach, we performed an evaluation using both synthetic and real event logs and process models. In particular, we tested it on a real-life case study about the management of road-traffic fines by an Italian town. The results show that our approach significantly improves the accuracy in determining probable explanations of nonconformity compared to existing techniques. Moreover, an analysis of the computation time shows the practical feasibility of our approach.

The paper is organized as follows. Section 2 introduces preliminary concepts. Section 3 provides the motivations for this work, discussing how the construction of optimal alignments should be kept independent of the reason why such alignments are constructed. Section 4 presents our approach for constructing optimal alignments. Section 5 presents experiment results, which are discussed in Sect. 6. Finally, Sect. 7 discusses related work and concludes the paper providing directions for future work.

2   Preliminaries

This section introduces the notation and preliminaries for our work.

2.1   Labeled Petri Nets, Event Logs, and Alignments

Process models describe how processes should be carried out. Many languages exist to model processes. Here, we use a simple formalism, which suffices for the purpose of this work:

Definition 1 (Labeled Petri Net). *A Labeled Petri net is a tuple* $(P, T, F, A, \ell, m_i, m_f)$ *where*

- *P is a set of places;*
- *T is a set of transitions;*
- $F \subseteq (P \times T) \cup (T \times P)$ *is the flow relation between places and transitions (and between transitions and places);*
- *A is the set of labels for transitions;*
- $\ell : T \rightarrow A$ *is a function that associates a label with every transition in T;*
- m_i *is the initial marking;*
- m_f *is the final marking.*

Hereafter, the simpler term Petri net is used to refer to Labeled Petri nets. The label of a transition identifies the activity represented by such a transition. Multiple transitions can be associated with the same activity label; this means that the same activity can be represented by multiple transitions. This is typically done to make the model simpler. Some transitions can be invisible. Invisible transitions do not correspond to actual activities but are necessary for routing purposes and, as such, their execution is never recorded in event logs. Given a Petri net N, $\mathsf{Inv}_N \subseteq A$ indicates the set of labels associated with invisible

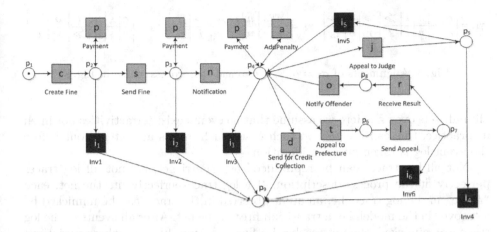

Fig. 1. A process model for managing road traffic fines. The green boxes represent the transitions that are associated with process activities while the black boxes represent invisible transitions. The text below the transitions represents the label, which is shortened with a single letter as indicated inside the transitions (Color figure online).

transitions. As a matter of fact, invisible transitions are also associated with labels, though these labels do not represent activities. We assume that a label associated with a visible transition cannot be also associated with invisible ones and vice versa.

The state of a Petri net is represented by a *marking*, i.e. a multiset of tokens on the places of the net. A Petri net has an initial marking m_i and a final marking m_f. When a transition is executed (i.e., *fired*), a token is taken from each of its input places and a token is added to each of its output places. A sequence of transitions σ_M leading from the initial to the final marking is a *complete process trace*. Given a Petri net N, Γ_N indicates the set of all complete process traces allowed by N.

Example 1. Figure 1 shows a normative process, expressed in terms of Petri net, which encodes the Italian laws and procedures to manage road traffic fines [19]. A process execution starts by recording a traffic fine in the system and sending it to Italian residents. Traffic fines might be paid before or after they are sent out by police or received by the offenders. Offenders are allowed to pay the due amount in partial payments. If the total amount of the fine is not paid in 180 days, a penalty is added. Offenders may appeal against a fine to the prefecture and/or judge. If an appeal is accepted, the fine management is closed. On the other hand, if the fine is not paid by the offender (and no appeal has been accepted), the process eventually terminates by handing over the case for credit collection.

Given a Petri net $N = (P, T, F, A, \ell, m_i, m_f)$, a log *trace* $\sigma_L \in A^*$ is a sequence of events where each event records the firing of a transition. In particular, each event records the label of the transition that has fired. An *event log* $\mathcal{L} \in \mathbb{B}(A)$ is a multiset of log traces, where $\mathbb{B}(X)$ is used to represent the set of

$$\gamma_1 = \left|\frac{c}{c}\left|\frac{s}{s}\right|\frac{n}{n}\left|\frac{t}{t}\right|\frac{\gg}{l}\left|\frac{\gg}{r}\right|\frac{o}{o}\left|\frac{\gg}{i_3}\right|\right. \quad \gamma_2 = \left|\frac{c}{c}\left|\frac{s}{s}\right|\frac{n}{n}\left|\frac{t}{t}\right|\frac{o}{\gg}\left|\frac{\gg}{l}\right|\frac{\gg}{i_6}\right| \quad \gamma_3 = \left|\frac{c}{c}\left|\frac{s}{s}\right|\frac{n}{n}\left|\frac{t}{\gg}\right|\frac{o}{\gg}\left|\frac{\gg}{d}\right|\right.$$

Fig. 2. Alignments of $\sigma_1 = \langle c, s, n, t, o \rangle$ and the process model in Fig. 1

all multisets over X. Here we assume that no events exist for activities not in A; in practice, this can happen: in such cases, such events are filtered out before the event log is taken into consideration.

Not all log traces can be reproduced by a Petri net, i.e. not all log traces perfectly fit the process description. If a log trace perfectly fits the net, each "move" in the log trace, i.e. an event observed in the trace, can be mimicked by a "move" in the model, i.e. a transition fired in the net. After all events in the log trace are mimicked, the net reaches its final marking. In cases where deviations occur, some moves in the log trace cannot be mimicked by the net or vice versa. We explicitly denote "no move" by $\gg$.

Definition 2 (Legal move). *Let $N = (P, T, F, A, \ell, m_i, m_f)$ be a Petri net. Let $S_L = (A \setminus Inv_N) \cup \{\gg\}$ and $S_M = A \cup \{\gg\}$. A legal move is a pair $(m_L, m_M) \in (S_L \times S_M) \setminus (\gg, \gg)$ such that*

- *(m_L, m_M) is a synchronous move if $m_L \in S_L$, $m_M \in S_M$ and $m_L = m_M$,*
- *(m_L, m_M) is a move on log if $m_L \in S_L$ and $m_M = \gg$,*
- *(m_L, m_M) is a move on model if $m_L = \gg$ and $m_M \in S_M$.*

Σ_N denotes the set of legal moves for a Petri net N.

In the remainder, we indicate that a sequence σ' is a prefix of a sequence σ'', denoted with $\sigma' \in (\sigma'')$, if there exists a sequence σ''' such that $\sigma'' = \sigma' \oplus \sigma'''$, where $\oplus$ denotes the concatenation operator.

Definition 3 (Alignment). *Let Σ_N be the set of legal moves for a Petri net $N = (P, T, F, A, \ell, m_i, m_f)$. An alignment of a log trace σ_L and N is a sequence $\gamma \in \Sigma_N^*$ such that, ignoring all occurrences of $\gg$, the projection on the first element yields σ_L and the projection on the second element yields a sequence $\langle a_1, \ldots, a_n \rangle$ such that there exists a sequence $\sigma'_P = \langle t_1, \ldots, t_n \rangle \in (\sigma_P)$ for some $\sigma_P \in \Gamma_N$ where, for each $1 \leq i \leq n$, $\ell(t_i) = a_i$. If $\sigma'_P \in \Gamma_N$, γ is called a complete alignment of σ_L and N.*

Figure 2 shows three possible complete alignments of a log trace $\sigma_1 = \langle c, s, n, t, o \rangle$ and the net in Fig. 1. The top row of an alignment shows the sequence of events in the log trace, and the bottom row shows the sequence of activities in the net (both ignoring $\gg$). Hereafter, we denote $|_L$ the projection of an alignment over the log trace and $|_P$ the projection over the net.

As shown in Fig. 2, there can be multiple possible alignments for a given log trace and process model. The quality of an alignment is measured based on a provided *cost function* $K : \Sigma_N^* \rightarrow \mathbb{R}_0^+$, which assigns a cost to each alignment

$\gamma \in \Sigma_N^*$. Typically, the cost of an alignment is defined as the sum of the costs of the individual moves in the alignment. An *optimal alignment* of a log trace and a process trace is one of the alignments with the lowest cost according to the provided cost function.

As an example, consider a cost function that assigns to any alignment a cost equal to the number of moves on log and model for visible transitions. If moves on model for invisible transitions i_k are ignored, γ_1 has two moves on model, γ_2 has one move on model and one move on log, and γ_3 has one move on model and two moves on log. Thus, according to the cost function, γ_1 and γ_2 are two optimal alignments of σ_1 and the process model in Fig. 1.

2.2   State Representation

At any point in time, a sequence of execution of activities leads to some state, and this state depends on which activities have been performed and in which order. Accordingly, any process execution can be mapped onto a state. As discussed in [3], a *state representation function* takes care of this mapping:

Definition 4 (State Representation). *Let A be a set of activity labels and R the set of possible state representations of the sequences in A^*. A state representation function* abst $: A^* \to R$ *produces a state representation* abst(σ) *for each process trace $\sigma \in \Gamma$.*

Several state representation functions can be defined. Each function leads to a different abstraction, meaning that multiple different traces can be mapped onto the same state, thus abstracting out certain trace's characteristics. Next, we provide some examples of state-representation functions:

Sequence abstraction. It is a trivial mapping where the abstraction preserves the order of activities. Each trace is mapped onto a state that is the trace itself, i.e. for each $\sigma \in A^*$, abst$(\sigma) = \sigma$.

Multi-set abstraction. The abstraction preserves the number of times each activity is executed. This means that, for each $\sigma \in A^*$, abst$(\sigma) = M \in \mathbb{B}(A)$ such that, for each $a \in A$, M contains all instances of a in σ.

Set abstraction. The abstraction preserves whether each activity has been executed or not. This means that, for each $\sigma \in A^*$, abst$(\sigma) = M \subseteq A$ such that, for each $a \in A$, M contains a if it ever occurs in σ.

Example 2. Table 1 shows the state representation of some process traces of the net in Fig. 1 using different abstractions. For instance, trace $\langle c, p, p, s, n \rangle$ can be represented as the trace itself using the sequence abstraction, as state $\{c(1), p(2), s(1), n(1)\}$ using the multi-set abstraction (in parenthesis the number of occurrences of activities in the trace), and as $\{c, p, s, n\}$ using the set abstraction. Traces $\langle c, p, s, n \rangle$ and $\langle c, p, p, s, n, p \rangle$ are also mapped to state $\{c, p, s, n\}$ using the set abstraction.

Table 1. Examples of state representation using different abstractions

Sequence	#	Multi-set	#	Set	#
$\langle c, p \rangle$	25	$\{c(1), p(1)\}$	25	$\{c, p\}$	25
$\langle c, s, n, p \rangle$	15	$\{c(1), p(1), s(1), n(1)\}$	15		
$\langle c, p, p, s, n \rangle$	5	$\{c(1), p(2), s(1), n(1)\}$	5	$\{c, p, s, n\}$	45
$\langle c, p, p, s, n, p \rangle$	25	$\{c(1), p(3), s(1), n(1)\}$	25		
$\langle c, s, n, a, d \rangle$	10	$\{c(1), s(1), n(1), a(1), d(1)\}$	10	$\{c, s, n, a, d\}$	10
$\langle c, s, n, p, a, d \rangle$	10	$\{c(1), s(1), n(1), p(1), a(1), d(1)\}$	10	$\{c, s, n, p, a, d\}$	10
$\langle c, s, n, p, t, l \rangle$	25	$\{c(1), s(1), n(1), p(1), t(1), l(1)\}$	30		
$\langle c, s, p, n, t, l \rangle$	5			$\{c, s, n, p, t, l\}$	60
$\langle c, p, s, n, p, t, l \rangle$	5	$\{c(1), s(1), n(1), p(2), t(1), l(1)\}$	30		
$\langle c, s, p, n, p, t, l \rangle$	25				
$\langle c, s, n, p, t, l, r, o \rangle$	50	$\{c(1), s(1), n(1), p(1), t(1), l(1), r(1), o(1)\}$	50	$\{c, s, n, p, t, l, r, o\}$	50

3   Constructing Optimal Alignments Is Purpose Independent

As discussed in Sect. 2.1, the quality of an alignment is determined with respect to a cost function. An optimal alignment provides the simplest and most parsimonious explanation with respect to the used cost function. Therefore, the choice of the cost function has a significant impact on the computation of optimal alignments.

Typically, process analysts define a cost function based on the context of use and the purpose of the analysis. For instance, Adriansyah et al. [7] study various ratios between the cost of moves on model and moves on log, and analyze their influence on the fitness of a trace with respect to a process model. The work in [5,6] uses alignments to identify nonconforming user behavior and quantify it with respect to a security perspective. In particular, the cost of deviations is determined in terms of which activity was executed, which user executed the activity along with its role, and which data have been accessed.

Existing alignment-based techniques make the implicit assumption that the obtained optimal alignments represent the most plausible explanations of what actually happened. However, they do not account for the fact that the use of different cost functions can yield different optimal alignments, thus resulting in inconsistent diagnostic information. The following example provides a concrete illustration of this issue.

Example 3. Consider the fine management process presented in Fig. 1 and the log trace $\sigma_2 = \langle c, s, a, d \rangle$. Suppose an analyst has to analyze σ_2 with respect to both fitness, in order to verify to what extent log traces comply with the behavior prescribed by the process model, and the information provided to citizens, in order to minimize the number of complaints and legal disputes. To this end, the analyst defines two cost functions, presented in Fig. 3a. Cost function c_1 defines the cost of deviations in terms of fitness. In particular, we use the cost

Moves	Cost Functions	
	c_1	c_2
$(p, \gg)$	5	1
$(\gg, p)$	1	1
$(a, \gg)$	5	1
$(\gg, a)$	1	1
$(s, \gg)$	5	1
$(\gg, s)$	1	5
$(n, \gg)$	5	1
$(\gg, n)$	1	5

(a) Cost Functions

c	s	$\gg$	a	d
c	s	n	a	d

c	s	a	d
c	s	$\gg$	$\gg$

(b) Optimal alignment using c_1 (c) Optimal alignment using c_2

Fig. 3. Inconsistent explanations of nonconformity due to the use of different cost functions.

function presented in [7] which defines a ratio between the cost of moves on log and the cost of moves on model equal to 5:1 for all activities. On the other hand, cost function c_2 defines the cost of deviations in terms of user satisfaction. Here, deviations concerning payment have low cost. On the other hand, the missed delivery of the fine or notification has a high cost. The optimal alignments obtained using cost functions c_1 and c_2 are given in Fig. 3b and c respectively.

Based on the example above, an interesting question comes up: which alignments should the analyst take as a plausible explanation of what happened? The alignments in Fig. 3b and c are supposed to be both plausible explanations, but with respect to different criteria. Our claim is that, although alignments provide a robust approach to conformance checking, it is necessary to rethink how cost functions are defined and, in general, how alignment-based techniques should be applied in practice.

This paper starts from the belief that the construction of an optimal alignment is independent from the purpose of the analysis. An optimal alignment should provide probable explanations of nonconformity, independently of why we are interested to know that. Therefore, first, an alignment providing probable explanations of what actually happened has to be constructed (hereafter we refer to such an alignment as *probable alignment*). Later, this alignment is analyzed according to the purpose of the analysis.

This separation of concerns can be achieved by employing two cost functions: a first cost function to find probable alignments and a second cost function to quantify the severity of the deviations of the computed alignments, which is customized according to the purpose of use. In the remainder of this paper, we discuss how to construct a cost function which provides probable explanations of what actually happened. In particular, this paper is concerned with constructing probable alignments; the discussion on the second purpose-dependent cost function is out of the scope of this paper.

4   History-Based Construction of Probable Alignments

This section presents our approach to construct alignments that give probable explanations of deviations based on objective facts, i.e. the historical logging data, rather than on subjective cost functions manually defined by process analysts. To construct an optimal alignment between a process model and an event log, we use the A-star algorithm [13], analogously to what proposed in [4].

Section 4.1 discusses how the cost of an alignment is computed, and Sect. 4.2 briefly reports on the use of A-star to compute probable alignments.

4.1   Definition of Cost Functions

The computation of probable alignments relies on a cost function that accounts for the probability of an activity to be executed in a certain state. The definition of such a cost function requires an analysis of the past history as recorded in the event log to compute the probability of an activity to immediately occur or to never eventually occur when the process execution is in a certain state.

The A-star algorithm [13] finds an optimal path from a source node to target node where optimal is defined in terms of minimal cost. In our context, moves that are associated to activities whose execution is more probable in a given state should have a low cost, whereas moves that are associated to activities whose execution is unlikely in a given state should have a high cost. Therefore, probabilities cannot be straightforwardly used as costs of moves. For this purpose, we need to introduce a class of functions $\mathcal{F} \subseteq [0, 1] \to \mathbb{R}^+$ to map probabilities to costs of moves. Based on the restriction imposed by the A-star algorithm on the choice of the cost function, a function $f \in \mathcal{F}$ if and only if $f(0) = \infty$ and f is monotonously decreasing between 0 and 1 (with $f(1) > 0$). Hereafter, these functions are called *cost profile*. Intuitively, a cost profile function is used to compute the cost of a legal move based on the probability that a given activity occurs when the process execution is in a given state. Below, we provide some examples of cost profile function:

$$f(p) = \tfrac{1}{p} \qquad f(p) = \tfrac{1}{\sqrt{p}} \qquad f(p) = 1 + \log\left(\tfrac{1}{p}\right) \tag{1}$$

The choice of the cost profile function has a significant impact on the computation of alignments (see Sect. 6). For instance, the first cost profile in Eq. 1 favorites alignments with more frequent traces, whereas the last cost profile is more sensitive to the number of deviations in the computed alignments. In Sect. 5, we evaluate these sample cost profiles with different combinations of event logs and process models. The purpose is to verify whether a cost profile universally works better than the others.

Similarly to what proposed in [4], the cost of an alignment move depends on the move type and the activity involved in the move. However, differently from [4], it also depends on the position in which the move is inserted:

Definition 5 (Cost of an alignment move). *Let Σ_N be the set of legal moves for a Petri net N. Let $\gamma \in \Sigma_N^*$ be a sequence of legal moves for N and $f \in \mathcal{F}$ a cost profile. The cost of appending a legal move $(m_L, m_M) \in \Sigma_N$ to γ with state-representation function* abst *is:*

$$\kappa_{\mathsf{abst}}((m_L, m_M), \gamma) =$$
$$\begin{cases} 0 & m_L = m_M \\ 0 & m_L = \gg \text{ and } m_M \in Inv_N \\ f\big(P_{\mathsf{abst}}(m_M \text{ occurs after } \gamma|_P)\big) & m_L = \gg \text{ and } m_M \notin Inv_N \\ f\big(P_{\mathsf{abst}}(m_L \text{ never eventually occurs after } \gamma|_P)\big) & m_M = \gg \end{cases}$$
$$(2)$$

Readers can observe that the cost of a move on log $(m_L, \gg)$ is not simply based on the probability of not executing activity m_L immediately after $\gamma|_P$; rather, it is based on the probability of never having activity m_M at the any moment in the future for that execution. This is motivated by the fact that a move on log $(m_L, \gg)$ indicates that m_L is not expected to ever occur in the future. Conversely, if it was expected, a number of moves in model would be introduced until the process model, modeled as a Petri net, reaches a marking that allows m_L to occur (and, thus, a move in both can be appended).

For a reliable computation of probabilities, we only use the subset of traces $\mathcal{L}_{fit}$ of the original event log $\mathcal{L}$ that fit the process model. We believe that, in many process analyses, it is not unrealistic to assume that several traces are compliant. For instance, this is the case for the real-life process about road-traffic fine management discussed in Sect. 5.2.

One may argue that some paths in the process model can be more prone to compliance errors compared to other paths. Thus, eliminating all non-fitting traces from the log would lead to underestimate the probability of executing activities in such paths. We argue that the reasons for nonconformity should be carefully investigated. For instance, frequent cases of nonconformity on a certain path may indicate that the process model does not reflect the reality [15, 24]. Ideally, an analyst should revise the process model and then use the new model to identify the set of fitting traces. This problem, however, is orthogonal to the current work and can be addressed using techniques for process repairing [14]. In this work, we assume that the process model is complete and accurately defines the business process. On the other hand, if the process model correctly reflects the reality, it is not obvious that non-fitting traces should be used to compute the cost function. Indeed, the resulting cost function would be biased by behavior that should not be permitted. Moreover, using error correction methods may lead to the problem of overfitting the training set [16]. Based on these considerations, we only use fitting traces as historical logging data.

The following two definitions describe how to compute the probabilities required by Definition 5.

Definition 6 (Probability that an activity occurs). *Let $\mathcal{L}$ be an event log and $\mathcal{L}_{fit} \subseteq \mathcal{L}$ the subset of traces that comply with a given process model represented by a Petri net $N = (P, T, F, A, \ell, m_i, m_f)$. The probability that an* *activity $a \in A$ occurs after executing σ with state-representation function* abst *is the ratio between number of traces in $\mathcal{L}_{fit}$ in which activity a is executed after reaching state* abst(σ) *and the total number of traces in $\mathcal{L}_{fit}$ that reach state* abst(σ):

$$P_{\mathsf{abst}}(a \text{ occurs after } \sigma) = \frac{|\{\sigma' \in \mathcal{L}_{fit} : \exists \sigma'' \in (\sigma'). \ \mathsf{abst}(\sigma'') = \mathsf{abst}(\sigma) \wedge \sigma'' \oplus \langle a \rangle \in (\sigma')\}|}{|\{\sigma' \in \mathcal{L}_{fit} : \exists \sigma'' \in (\sigma'). \ \mathsf{abst}(\sigma'') = \mathsf{abst}(\sigma)\}|} \quad (3)$$

Definition 7 (Probability that an activity never eventually occurs).
Let $\mathcal{L}$ be an event log and $\mathcal{L}_{fit} \subseteq \mathcal{L}$ the subset of traces that comply with a given process model represented by a Petri net $N = (P, T, F, A, \ell, m_i, m_f)$. The probability that an activity $a \in A$ will never eventually occur in a process execution after executing $\sigma \in A^*$ with state-representation function abst *is the ratio between the number of traces in $\mathcal{L}_{fit}$ in which a is never eventually executed after reaching state* abst(σ) *and the total number of traces in $\mathcal{L}_{fit}$ that reach state* abst(σ):

$$P_{\mathsf{abst}}(a \text{ never eventually occurs after } \sigma) =$$
$$\frac{|\{\sigma' \in \mathcal{L}_{fit} : \exists \sigma'' \in (\sigma'). \ \mathsf{abst}(\sigma'') = \mathsf{abst}(\sigma) \wedge \forall \sigma''' \ \sigma'' \oplus \sigma''' \oplus \langle a' \rangle \in (\sigma') \wedge a' \neq a\}|}{|\{\sigma' \in \mathcal{L}_{fit} : \exists \sigma'' \in (\sigma'). \ \mathsf{abst}(\sigma'') = \mathsf{abst}(\sigma)\}|} \quad (4)$$

Intuitively, $P_{\mathsf{abst}}(a \text{ occurs after } \sigma)$ and $P_{\mathsf{abst}}(a \text{ never eventually occurs after } \sigma)$ are conditional probabilities. Given two events A and B, the conditional probability of A given B is defined as the quotient of the probability of the conjunction of events A and B, and the probability of B:

$$P(A|B) = \frac{P(A \cap B)}{P(B)} \quad (5)$$

It is easy to verify that Eq. 3 coincides with Eq. 5 where A represents that activity a is executed, B that trace σ is executed, and $A \cap B$ that $\sigma \oplus \langle a \rangle$ is executed. Similar observations hold for Eq. 4.

The cost of an alignment is the sum of the cost of all moves in the alignment, which are computed as described in Definition 5:

Definition 8 (Cost of an alignment). *Let Σ_N be the set of legal moves for a Petri net N. The cost of alignment $\gamma \in \Sigma_N^*$ with state-representation function* abst *is computed as follows:*

$$K_{\mathsf{abst}}(\gamma \oplus (m_L, m_M)) = \begin{cases} \kappa_{\mathsf{abst}}((m_L, m_M), \langle \rangle) & \gamma = \langle \rangle \\ \kappa_{\mathsf{abst}}((m_L, m_M), \gamma) + K_{\mathsf{abst}}(\gamma) & otherwise \end{cases} \quad (6)$$

Hereafter, the term *probable alignment* is used to denote any of the optimal alignments (i.e., alignments with the lowest cost) according to the cost function given in Definition 8.

4.2 The Use of the A-Star Algorithm to Construct Alignments

The A-star algorithm [13] aims to find a path in a graph V from a given *source* node v_0 to any node $v \in V$ in a target set. Every node v of graph V is associated with a cost determined by an *evaluation* function $f(v) = g(v) + h(v)$, where

- $g : V \to \mathbb{R}_0^+$ is a function that returns the cost of the smallest path from v_0 to v;
- $h : V \to \mathbb{R}_0^+$ is a heuristic function that estimates the cost of the path from v to its preferred target node.

Function h is said to be *admissible* if it returns a value that underestimates the distance of a path from a node v' to its preferred target node v'', i.e. $g(v') + h(v') \leq g(v'')$. If h is admissible, A-star finds a path that is guaranteed to have the overall lowest cost.

The A-star algorithm keeps a priority queue of nodes to be visited: higher priority is given to nodes with lower costs. The algorithm works iteratively: at each step, the node v with lowest cost is taken from the priority queue. If v belongs to the target set, the algorithm ends returning node v. Otherwise, v is expanded: every successor v' is added to the priority queue with a cost $f(v')$.

We employ A-star to find any of the optimal alignments between a log trace $\sigma_L \in \mathcal{L}$ and a Petri net N. In order to be able to apply A-star, an opportune search space needs to be defined. Every node γ of the search space V is associated to a different alignment that is a prefix of some complete alignment of σ_L and N. Since a different alignment is also associated to every search-space node and vice versa, we use the alignment to refer to the associated state. The source node is an empty alignment $\gamma_0 = \langle \rangle$ and the set of target nodes includes every complete alignment of σ_L and N.

Let us denote the length of a sequence σ with $\|\sigma\|$. Given a node/alignment $\gamma \in V$, the search-space successors of γ include all alignments $\gamma' \in V$ obtained from γ by concatenating exactly one move. Given an alignment $\gamma \in V$, the cost of the path from the initial node to node $\gamma \in V$ is:

$$g(\gamma) = \|\gamma|_L\| + K(\gamma).$$

where $K(\gamma)$ is the cost of alignment γ according to Definition 8. It is easy to check that, given two complete alignments γ'_C and γ''_C, $K(\gamma'_C) < K(\gamma''_C)$ iff $g(\gamma'_C) < g(\gamma''_C)$ and $K(\gamma'_C) = K(\gamma''_C)$ iff $g(\gamma'_C) = g(\gamma''_C)$. Therefore, an optimal solution returned by A-star coincides with an optimal alignment.

The time complexity of A-star depends on the heuristic used to find an optimal solution. In this work, we consider term $\|\sigma_L\|$ in h to define an admissible heuristic; this term does not affect the optimality of solutions. Given an alignment $\gamma \in V$, we employ the heuristics:

$$h(\gamma) = \|\sigma_L\| - \|\gamma|_L\|.$$

For alignment γ, the number of steps to add in order to reach a complete alignment is lower bounded by the number of execution steps of trace σ_L that have not been included yet in the alignment, i.e. $\|\sigma_L\| - \|\gamma|_L\|$. Since the additional cost to traverse a single node is at least 1, the cost to reach a target node is at least $h(\gamma)$, corresponding to the case where the part of the log trace that still needs to be included in the alignment perfectly fits.

$$\gamma' = \underbrace{\begin{array}{|c|c|c|} \hline c & s & n \\ \hline c & s & n \\ \hline \end{array}}_{\gamma} \quad \oplus \quad \left\{ \begin{array}{ll} (l, \gg) & \kappa\big((l, \gg), \gamma\big) = 1.49 \\ (\gg, p) & \kappa\big((\gg, p), \gamma\big) = 1.04 \\ (\gg, a) & \kappa\big((\gg, a), \gamma\big) = 2.04 \\ (\gg, d) & \kappa\big((\gg, d), \gamma\big) = \infty \\ & \cdots \end{array} \right.$$

Fig. 4. Construction of the alignment of log trace $\sigma_3 = \langle c, s, n, l, o \rangle$ and the net in Fig. 1. Cost of moves are computed with sequence state-representation function, cost profile $f(p) = 1 + \log(1/p)$, and $\mathcal{L}_{fit}$ in Table 1.

Example 4. Consider a log trace $\sigma_3 = \langle c, s, n, l, o \rangle$ and the net N in Fig. 1. An analyst wants to determine probable explanations of nonconformity by constructing probable alignments of σ_3 and N, based on historical logging data. In particular, $\mathcal{L}_{fit}$ consists of the traces in Table 1 (the first column shows the traces, and the second the number of occurrences of a trace in the history). Assume that the A-star algorithm has constructed an optimal alignment γ of trace $\langle c, s, n \rangle \in (\sigma_3)$ and N (left part of Fig. 4). The next event in the log trace (i.e., l) cannot be replayed in the net. Therefore, the algorithm should determine which move is the most likely to have occurred. Different moves are possible; for instance, a move on log for l, a move on model for p, a move on model for t, etc. The algorithm computes the cost for these moves using Eq. 5 (right part of Fig. 4). As move on model $(\gg, p)$ is the move with the least cost (and no other alignments have lower cost), alignment $\gamma' = \gamma \oplus (\gg, p)$ is selected for the next iteration. It is worth noting that activity d never occurs after $\langle c, s, n \rangle$ in $\mathcal{L}_{fit}$; consequently, the cost of move $(\gg, d)$ is equal to ∞.

5   Implementation and Experiments

We have implemented our approach for history-based construction of alignments as a plug-in of the nightly-build version of the ProM framework (http://www.promtools.org/prom6/nightly/).[1] The plug-in takes as input a process model and two event logs. It computes probable alignments for each trace in the first event log with respect to the process model based on the frequency of the traces in the second event log (historical logging data). The output of the plug-in is a set of alignments and can be used by other plug-ins for further analysis. A screenshot of the plugin is shown in Fig. 5. In particular, the figure shows the result of aligning a few sample event traces with the net in Fig. 1.

To assess the practical feasibility and accuracy of the approach, we performed a number of experiments using both synthetic and real-life logs. In the experiments with synthetic logs, we assumed that the execution of an activity depends on the activities that were performed in the past. In the experiments with real-life logs, we tested if this assumption holds in real applications. Accordingly, the

---

[1] The plug-in is available in package *History-Based Conformance Checking*.

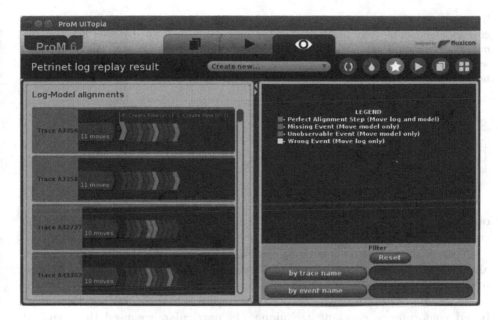

Fig. 5. Screenshot of the implemented approach in ProM, showing the probable alignment constructed between log traces and the process model in Fig. 1.

real-life logs were used as historical logging data. To evaluate the approach, we artificially added noise to the traces used for the experiments. This was necessary to assess the ability of the approach to reconstruct the original traces. The experiments were performed using a machine with 3.4 GHz Intel Core i7 processor and 16 GB of memory.

5.1 Synthetic Data

For the experiments with synthetic data, we used the process for handling credit requests in [19]. Based on this model, we generated 10000 traces consisting of 69504 events using the CPN Tools (http://cpntools.org). To assess the accuracy of the approach, we manipulated 20 % of these traces by introducing different percentages of noise. In particular, given a trace, we added and removed a number of activities to/from the trace equal to the same percentage of the trace length. The other traces were used as historical logging data. We computed probable alignments of the manipulated traces and process model, and evaluated the ability of the approach to reconstruct the original traces. To this end, we measured the percentage of correct alignments (i.e., the cases where a projection of an alignment over the process coincides with the original trace) and compute the overall Levenshtein distance [17] between the original traces and the projection of the computed alignments over the process. The Levenshtein distance is a string metric that measures the distance between two sequences, i.e. the minimal number of changes required to transform one sequence into the

Table 2. Results of experiments on synthetic data. CA indicates the percentage of correct alignments, and LD indicates the overall Levenshtein distance between the original traces and the projection of the alignments over the process. For comparison with existing approaches, the standard cost function as defined in [4] was used. The best results for each amount of noise are highlighted in bold.

| | $1/p$ | | | | | | $1/\sqrt{p}$ | | | | | | $1 + \log(1/p)$ | | | | | | Existing approach | |
| | Seq | | Multi-set | | Set | | Seq | | Multi-set | | Set | | Seq | | Multi-set | | Set | | | |
Noise	CA	LD	CA	LD	CA	LD	CA	LD	CA	LD	CA	LD	CA	LD	CA	LD	CA	LD	CA	LD
10%	93	259	93	258	87	514	**95**	164	**95**	164	88	430	**95**	**153**	**95**	154	88	409	92	233
20%	85	569	85	561	78	968	**87**	426	**87**	431	79	852	**87**	**410**	**87**	415	79	823	83	534
30%	74	1084	74	1077	65	1653	**76**	950	75	963	66	1509	**76**	**944**	75	958	67	1474	71	1110
40%	63	1658	62	1659	55	2285	**64**	1519	**64**	1537	56	2148	**64**	**1512**	**64**	1535	56	2118	60	1685

other. In our setting, it provides an indication of how much the projection of the computed alignments over the process is close to the original traces.

We tested our approach with different amounts of noise (i.e., 10 %, 20 %, 30 % and 40 % of the trace length), with different cost profiles (i.e., $1/p$, $1/\sqrt{p}$, and $1 + \log(1/p)$), and with different state-representation functions (i.e., *sequence*, *multiset*, and *set*). Moreover, we compared our approach with existing alignment-based conformance checking techniques. In particular, we used the standard cost function introduced in [4]. We repeated each experiment five times. Table 2 shows the results where every entry reports the average over the five runs.

The results show that cost profiles $1/\sqrt{p}$ and $1 + \log(1/p)$ in combination with *sequence* and *multi-set* abstractions are able to better identify what really happened, i.e. they align the manipulated traces with the corresponding original traces in more cases (CA). In all cases, cost profile $1 + \log(1/p)$ with *sequence* state-representation function provides more accurate diagnostics (LD): even if log traces are not aligned to the original traces, the projection over the process of alignments constructed using this cost profile and abstraction are closer to the original traces. Compared to the cost function used in [4], our approach computed the correct alignment for 4.4 % more traces when cost profile $1 + \log(1/p)$ and *sequence* state-representation function are used. In particular, our approach correctly reconstructed the original trace for 18.4 % of the traces that were not correctly reconstructed using the cost function used in [4]. Moreover, an analysis of LD shows that, on average, the traces reconstructed using our approach have 0.37 deviations (compared to the original traces), while the traces reconstructed using the cost function used in [4] have 0.45 deviation. This corresponds to an improvement of LD of about 15.2 %.

5.2 Real-Life Logs

To evaluate the applicability of our approach to real-life scenarios, we used an event log obtained from a fine management system of the Italian police [19].[2] The process model in form of Petri net is presented in Fig. 1. We extracted a log

---

[2] The event log is also available for download: http://dx.doi.org/10.4121/uuid: 270fd440-1057-4fb9-89a9-b699b47990f5.

Table 3. Results of experiments on real-life data. Notation analogous to Table 2.

	1/p						1/√p						1 + log(1/p)						Existing approach	
	Seq		Multi-set		Set		Seq		Multi-set		Set		Seq		Multi-set		Set			
Noise	CA	LD	CA	LD	CA	LD	CA	LD	CA	LD	CA	LD	CA	LD	CA	LD	CA	LD	CA	LD
10%	99	397	99	397	99	415	99	384	99	389	99	408	99	**366**	99	371	99	389	98	1274
20%	99	585	99	585	99	602	99	570	99	575	99	592	99	**554**	99	559	99	576	97	1448
30%	89	3349	89	3349	89	3371	89	3300	89	3341	89	3362	89	**3281**	89	3322	89	3344	87	4284
40%	76	9160	76	9160	75	9238	**76**	**9091**	76	9152	75	9230	76	9103	75	9165	75	9243	74	9861

consisting of 142408 traces and 527549 events, where all traces are conforming to the net. To these traces, we applied the same methodology used for the experiments reported in Sect. 5.1. We repeated the experiments five times. Table 3 shows the results where every entry reports the average over the five runs.

The results confirm that cost profiles $1/\sqrt{p}$ and $1+log(1/p)$ in combination with *sequence* and *multi-set* state-representation functions provide the more accurate diagnostics (both CA and LD). Moreover, the results show that our approach (regardless of the used cost profile and state-representation function) performs better than the cost function in [4] on real-life logs. In particular, using *sequence* state-representation function and cost profile $1 + log(1/p)$, our approaches computed the correct alignment for 1.8 % more traces than what the cost function in [4] did. Although this may not be seen as a significant improvement, it is worth noting that the cost function in [4] already reconstructs most of the traces (98 % and 97 % of the traces for 10 % and 20 % noise respectively). Nonetheless, our approach correctly reconstructed the original trace for 19.3 % of the traces that were not correctly reconstructed using the cost function used in [4]. Moreover, our approach improves LD by 21.1 % compared to the cost function used in [4]. Such an improvement shows that when the original trace is not reconstructed correctly, our approach returns an explanation that is significantly closer to what actually happened.

5.3 Complexity Analysis

In the previous sections, we have analyzed the accuracy of our approach for the computation of probable alignments. In this section, we aim to perform a complexity analysis. In the worst case, the problem is clearly exponential in the length of the log traces and the number of process activities. However, in this paper, we advocate the use of the A-star algorithm since it can drastically reduce the execution time in the average case. To illustrate this, we report on the computation time for the loan process and the fine-management process for different amounts of noise.

Figure 6 shows the distribution of the computation time for the traces used in the experiments. In particular, Fig. 6a shows that, in the experiments of Sect. 5.1 (loan process), the construction of alignments required less than 1 ms for most of the traces. On the other hand, the construction of probable alignments for the fine management process required less than 0.3 ms for most of the traces (Fig. 6b). Table 4 reports the mean and standard deviation of computation time

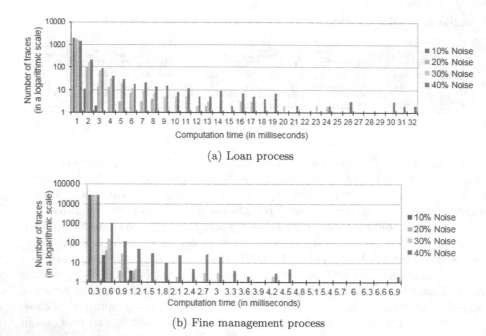

(a) Loan process

(b) Fine management process

Fig. 6. Distribution of the computation time required to construct probable alignments for different amounts of noise. The computation time is grouped into 1 ms intervals in Fig. 6a and 0.3 ms intervals in Fig. 6b. The y-axis values are shown in a logarithmic scale.

Table 4. Mean and standard deviation of computation time required to construct probable alignments for different amounts of noise.

<table>
<tr><td colspan="3">(a) Loan process</td><td colspan="3">(b) Fine management process</td></tr>
<tr><th>Noise</th><th>Mean</th><th>Standard Deviation</th><th>Noise</th><th>Mean</th><th>Standard Deviation</th></tr>
<tr><td>10%</td><td>0.255</td><td>0.635</td><td>10%</td><td>0.102</td><td>0.042</td></tr>
<tr><td>20%</td><td>0.421</td><td>0.935</td><td>20%</td><td>0.111</td><td>0.047</td></tr>
<tr><td>30%</td><td>0.999</td><td>3.280</td><td>30%</td><td>0.110</td><td>0.091</td></tr>
<tr><td>40%</td><td>3.014</td><td>14.146</td><td>40%</td><td>0.139</td><td>0.232</td></tr>
</table>

required to construct probable alignments for different levels of noise. The results show that, in both experiments, the time needed to construct probable alignments increases with increasing amounts of noise.

Based on the results presented in this section, we can conclude that, for both synthetic and real-life processes, our approach can construct probable alignments for a trace in the order of magnitude of milliseconds, which shows its practical feasibility.

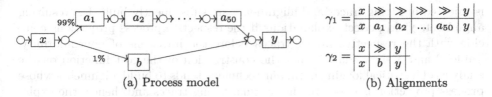

(a) Process model		(b) Alignments

Fig. 7. Process model including two paths formed by a (sub)sequence of 50 activities and 1 activity respectively. The first path is executed in 99 % of the cases; the second in 1 % of the cases. γ_1 and γ_2 are two possible alignments of trace $\sigma = \langle x, y \rangle$ and the process model.

6 Discussion

The A-star algorithm requires a cost function to penalize nonconformity. In our experiments, we have considered a number of cost profiles to compute the cost of moves on log/model based on the probability of a given activity to occur in historical logging data. The selection of the cost profile has a significant impact on the results as they penalize deviations differently. For instance, cost profile $1/p$ penalizes less probable moves much more than $1 + log(1/p)$. To illustrate this, consider a trace $\sigma = \langle x, y \rangle$ and the process model in Fig. 7a. Two possible alignments, namely γ_1 and γ_2, are conceivable (Fig. 7b). γ_1 contains a large number of deviations compared to γ_2 (50 moves on log vs. 1 move on log). The use of cost profile $1/p$ yields γ_1 as the optimal alignment, while the use of cost profile $1 + log(1/p)$ yields γ_2 as the optimal alignment. Tables 2 and 3 show that cost profile $1 + log(1/p)$ usually provides more accurate results. Cost profile $1/p$ penalizes less probable moves excessively, and thus tends to construct alignments with more frequent traces in the historical logging data even if those alignments contain a significantly larger number of deviations. Our experiments suggest that the construction of probable alignments requires a trade-off between the frequency of the traces in historical logging data and the number of deviations in alignments, which is better captured by cost profile $1 + log(1/p)$.

Different state-representation functions can be used to characterize the state of a process execution. In this work, we have considered three state-representation functions: *sequence*, *multi-set*, and *set*. The experiments show that in general the sequence abstraction produces more accurate results compared to the other abstractions. The set abstraction provides the least accurate results, especially when applied to the process for handling credit requests (Table 2). The main reason is that this abstraction is not able to accurately characterize the state of the process, especially in presence of loops: after each loop iteration the process execution yields the same state. Therefore, the cost function constructed using the set abstraction is not able to account for the fact that the probability of executing certain activities can increase after every loop iteration, thus leading to alignments in which loops are not captured properly.

The experiments show that our technique tends to build alignments that provide better explanations of deviations. It is easy to see that, when nonconformity is injected in fitting traces and alignments are subsequently built, the resulting alignments yield perfect explanations if the respective process projections coincide with the respective fitting traces before the injections of nonconformity. Tables 2 and 3 show that, basing the construction of the cost function on the analysis of historical logging data, our technique tends to build alignments whose process projection is closer to the original fitting traces and, hence, the explanations of deviations are closer to the correct ones.

7   Related Work and Conclusions

In process mining, a number of approaches have been proposed to check conformance of process models and the actual behavior recorded in event logs. Some approaches [10,11,18,21,22] check conformance by verifying whether traces satisfies rules encoding properties expected from the process. Petković et al. [23] verify whether a log trace is a valid trace of the transition system generated by the process model. Rozinat et al. [24] propose a token-based approach for checking conformance of an event log and a Petri net. The number of missing and added tokens after replaying traces is used to measure the conformance between the log and the net. Banescu et al. [9] extend the work in [24] to identify and classify deviations by analyzing the configuration of missing and added tokens using deviation patterns. The genetic mining algorithm in [20] uses a similar replay technique to measure the quality of process models with respect to given executions. However, these approaches only give a Boolean answers diagnosing whether traces conform to a process model or not. When they are able to provide diagnostic information, such information is often imprecise. For instance, token-based approaches may allow behavior that is not allowed by the model due to the used heuristics and thus may provide incorrect diagnostic information.

Recently, the construction of alignments has been proposed as a robust approach for checking the conformance of event logs with a given process model [4]. Alignments have proven to be powerful artifacts to perform conformance checking. By constructing alignments, analysts can be provided with richer and more accurate diagnostic information. In fact, alignments are also used as the main enablers for a number of techniques for process analytics, auditing, and process improvement, such as for performance analysis [2], privacy compliance [5,6] and process-model repairing [14].

To our knowledge, the main problem of existing techniques for constructing optimal alignments is related to the fact that process analysts need to provide a function which associates a cost to every possible deviation. These cost functions are only based on human judgment and, hence, prone to imperfections. If alignment-based techniques are fed with imprecise cost functions, they create imperfect alignments, which ultimately leads to unlikely or, even, incorrect diagnostics.

In this paper, we have proposed a different approach where the cost function is automatically computed based on real facts: historical logging data recorded in event logs. In particular, the cost function is computed based on the probability of activities to be executed or not in a certain state (representing which activities have been executed and their order). Experiments have shown that, indeed, our approach can provide more accurate explanations of nonconformity of process executions, if compared with existing techniques.

We acknowledge that the evaluation is far from being completed. We aim to perform more extensive experiments to verify whether certain cost-profile functions provide more probable alignments than others or, at least, to give some guidelines to determine in which settings a given cost-profile function is preferable.

In this paper, we only considered the control flow, i.e. the name of the activities and their ordering, to construct the cost function and, hence, to compute probable alignments. However, the choice in a process execution is often driven by other aspects. For instance, when instances are running late, the execution of certain fast activities are more probable; or, if a certain process attribute takes on a given value, certain activities are more likely to be executed. We expect that our approach can be significantly improved if other business process perspectives (e.g., data, time and resources) are taken into account.

Acknowledgement. This work has been partially funded by the NWO CyberSecurity programme under the PriCE project, the Dutch national program COMMIT under the THeCS project and the European Communitys Seventh Framework Program FP7 under grant agreement no. 603993 (CORE).

References

1. van der Aalst, W.M.P.: Process Mining - Discovery, Conformance and Enhancement of Business Processes. Springer, Heidelberg (2011)
2. van der Aalst, W.M.P., Adriansyah, A., van Dongen, B.F.: Replaying history on process models for conformance checking and performance analysis. Data Min. Knowl. Discov. **2**(2), 182–192 (2012)
3. van der Aalst, W.M.P., Schonenberg, M.H., Song, M.: Time prediction based on process mining. Inf. Syst. **36**(2), 450–475 (2011)
4. Adriansyah, A., van Dongen, B.F., van der Aalst, W.M.P.: Memory-efficient alignment of observed and modeled behavior. BPM Center Report 03–03, BPMcenter.org (2013)
5. Adriansyah, A., van Dongen, B.F., Zannone, N.: Controlling break-the-glass through alignment. In: Proceedings of International Conference on Social Computing. pp. 606–611. IEEE (2013)
6. Adriansyah, A., van Dongen, B.F., Zannone, N.: Privacy analysis of user behavior using alignments. it - Inf. Technol. **55**(6), 255–260 (2013)
7. Adriansyah, A., Sidorova, N., van Dongen, B.F.: Cost-based fitness in conformance checking. In: Proceedings of International Conference on Application of Concurrency to System Design. pp. 57–66. IEEE (2011)
8. Alizadeh, M., de Leoni, M., Zannone, N.: History-based construction of log-process alignments for conformance checking: discovering what really went wrong. In: Proceedings of the 4th International Symposium on Data-Driven Process Discovery and Analysis. CEUR Workshop Proceedings, vol. 1293, pp. 1–15. CEUR-WS.org (2014)

9. Banescu, S., Petković, M., Zannone, N.: Measuring Privacy Compliance Using Fitness Metrics. In: Barros, A., Gal, A., Kindler, E. (eds.) BPM 2012. LNCS, vol. 7481, pp. 114–119. Springer, Heidelberg (2012)
10. Borrego, D., Barba, I.: Conformance checking and diagnosis for declarative business process models in data-aware scenarios. Expert Syst. Appl. **41**(11), 5340–5352 (2014)
11. Caron, F., Vanthienen, J., Baesens, B.: Comprehensive rule-based compliance checking and risk management with process mining. Decis. Support Syst. **54**(3), 1357–1369 (2013)
12. Cook, J.E., Wolf, A.L.: Software process validation: quantitatively measuring the correspondence of a process to a model. TOSEM **8**(2), 147–176 (1999)
13. Dechter, R., Pearl, J.: Generalized best-first search strategies and the optimality of A*. J. ACM **32**, 505–536 (1985)
14. Fahland, D., van der Aalst, W.M.P.: Model repair - aligning process models to reality. Inf. Syst. **47**, 220–243 (2014)
15. Fernández-Ropero, M., Reijers, H.A., Pérez-Castillo, R., Piattini, M.: Repairing business process models as retrieved from source code. In: Nurcan, S., Proper, H.A., Soffer, P., Krogstie, J., Schmidt, R., Halpin, T., Bider, I. (eds.) BPMDS 2013 and EMMSAD 2013. LNBIP, vol. 147, pp. 94–108. Springer, Heidelberg (2013)
16. Geman, S., Bienenstock, E., Doursat, R.: Neural networks and the bias/variance dilemma. Neural Comput. **4**(1), 1–58 (1992)
17. Levenshtein, V.: Binary codes capable of correcting deletions, insertions, and reversals. Sov. Phys. Dokl. **10**(8), 707–710 (1966)
18. Ly, L.T., Rinderle-Ma, S., Göser, K., Dadam, P.: On enabling integrated process compliance with semantic constraints in process management systems - requirements, challenges, solutions. Inf. Syst. Front. **14**(2), 195–219 (2012)
19. Mannhardt, F., de Leoni, M., van der Aalst, W.M.P.: Balanced multi-perspective checking of process conformance. BPM Center Report 14–08, BPMcenter.org (2014)
20. de Medeiros, A.K.A., Weijters, A., van der Aalst, W.M.P.: Genetic process mining: an experimental evaluation. Data Min. Knowl. Discov. **14**(2), 245–304 (2007)
21. de Medeiros, A.K.A., van der Aalst. Wil M. P., Pedrinaci, C.: Semantic process mining tools: core building blocks. In: Proceedings of ECIS, pp. 1953–1964. AIS (2008)
22. Montali, M.: Declarative open interaction models. In: Montali, M. (ed.) Specification and Verification of Declarative Open Interaction Models. LNBIP, vol. 56, pp. 11–45. Springer, Heidelberg (2010)
23. Petković, M., Prandi, D., Zannone, N.: Purpose control: did you process the data for the intended purpose? In: Jonker, W., Petković, M. (eds.) SDM 2011. LNCS, vol. 6933, pp. 145–168. Springer, Heidelberg (2011)
24. Rozinat, A., van der Aalst, W.M.P.: Conformance checking of processes based on monitoring real behavior. Inf. Syst. **33**(1), 64–95 (2008)

Dynamic Constructs Competition Miner - Occurrence- vs. Time-Based Ageing

David Redlich[1,2(⊠)], Thomas Molka[1,3], Wasif Gilani[1], Gordon Blair[2], and Awais Rashid[2]

[1] SAP Research Center Belfast, Belfast, UK
{david.redlich,thomas.molka,wasif.gilani}@sap.com
[2] Lancaster University, Lancaster, UK
{gordon,marash}@comp.lancs.ac.uk
[3] University of Manchester, Manchester, UK

Abstract. Since the environment for businesses is becoming more competitive by the day, business organizations have to be more adaptive to environmental changes and are constantly in a process of optimization. Fundamental parts of these organizations are their business processes. Discovering and understanding the actual execution flow of the processes deployed in organizations is an important enabler for the management, analysis, and optimization of both, the processes and the business. This has become increasingly difficult since business processes are now often dynamically changing and may produce hundreds of events per second. The basis for this paper is the Constructs Competition Miner (CCM). A divide-and-conquer algorithm which discovers block-structured processes from event logs possibly consisting of exceptional behaviour. In this paper we propose a set of modifications for the CCM to enable dynamic business process discovery of a run-time process model from a stream of events. We describe the different modifications with a particular focus on the influence of individual events, i.e. ageing techniques. We furthermore investigate the behaviour and performance of the algorithm and the ageing techniques on event streams of dynamically changing processes.

Keywords: Run-time models · Business Process Management · Process mining · Complex Event Processing · Event streaming · Big Data

1 Introduction

The success of modern organizations has become increasingly dependent on the efficiency and performance of their employed business processes (BPs). These processes dictate the execution order of singular tasks to achieve certain business goals and hence represent fundamental parts of most organizations. In the context of business process management, the recent emergence of Big Data yields new challenges, e.g. more analytical possibilities but also additional run-time constraints. An important discipline in this area is Process Discovery: It is concerned

© IFIP International Federation for Information Processing 2015
P. Ceravolo et al. (Eds.): SIMPDA 2014, LNBIP 237, pp. 79–106, 2015.
DOI: 10.1007/978-3-319-27243-6_4

with deriving process-related information from event logs and, thus, enabling the business analyst to extract and understand the actual behaviour of a business process. Even though they are now increasingly used in commercial settings, many of the developed process discovery algorithms were designed to work in a static fashion, e.g. as provided by the ProM framework [21], but are not easily applicable for processing real-time event streams. Additionally, the emergence of Big Data results in a new set of challenges for process discovery on event streams, for instance [16,22]: (1) *high event frequency* (e.g. thousands of events per second), and (2) *less rigid processes* (e.g. BPs found on the operational level of e-Health and security use-cases are usually subjected to frequent changes).

In order to address the challenges we propose modifications for the Constructs Competition Miner (CCM) [15] to enable dynamic process discovery as proposed in [16]. The CCM is a process discovery algorithm that follows a divide-and-conquer approach to directly mine a block-structured process model which consists of common BP-domain constructs and represents the main behaviour of the process. This is achieved by calculating global relations between activities and letting different *constructs* compete with each other for the most suitable solution from top to bottom using "soft" constraints and behaviour approximations. The CCM was designed to deal with noise and not-supported behaviour. To apply the CCM on event streams the algorithm was split up into two individually operating parts:

1. **Run-Time Footprint Calculation**, i.e. the current footprint[1], which represents the abstract "state" of the system, is updated with occurrence of each event. The influence of individual events on the run-time footprint is determined by two different strategies: time-based and occurrence-based ageing. Since every occurring event constitutes a system state transition, the algorithmic execution-time needs to be kept to a minimum.
2. **Scheduled Footprint Interpretation**, i.e. from the footprint the current business process is discovered in a scheduled, reoccurring fashion. Since this part is executed in a different lifecycle it has less execution-time constraints. In this step the abstract "computer-centric" footprint is transformed into a "human-centric" business process representation.

The remainder of this paper provides essential background information (Sect. 2), a discussion of related work (Sect. 3), a summarized description of the original CCM (Sect. 4), the modifications that were carried out on top of the CCM to enable Scalable Dynamic Process Discovery (Sect. 5), an evaluation of the behaviour of the resulting algorithm for event streams of dynamically changing processes (Sect. 6), and an outlook of future work (Sect. 7).

2   Background

Business Processes are an integral part of modern organizations, describing the set of activities that need to be performed, their order of execution, and the

---

[1] footprint is a term used in the process discovery domain, abstractly representing existent "behaviour" of a log, e.g. activity "a" is followed by activity "b".

entities that execute them. Prominent BP examples are Order-to-Cash or Procure-to-Pay. According to Ko et al. BPs are defined as *"...a series or network of value-added activities, performed by their relevant roles or collaborators, to purposefully achieve the common business goal"* [8]. A BP is usually described by a *process model* conforming to a business process standard, e.g. Business Process Model and Notation (BPMN) [14], or Yet Another Workflow Language (YAWL) [19]. In this paper, we will focus on business processes consisting of a set of common control-flow elements, supported by most of the existing BP standards: start and end events, activities (i.e. process steps), parallel gateways (AND-Split/Join), and exclusive gateways (XOR-Split/Join) (see [14,19]). In Fig. 1 an example process involving all the introduced elements is displayed. Formally, we define a business process model as follows [15]:

Definition 1. *A business process model is a tupel* $BP = (A, S, J, E_s, E_e, C)$ *where A is a finite set of activities, S a finite set of splits, J a finite set of joins, E_s a finite set of start events, E_e a finite set of end events, and $C \subseteq F \times F$ the path connection relation, with $F = A \cup S \cup J \cup E_s \cup E_e$, such that*

- $C = \{(c_1, c_2) \in F \times F \mid c_1 \neq c_2 \wedge c_1 \notin E_e \wedge c_2 \notin E_s\}$,
- $\forall a \in A \cup J \cup E_s : |\{(a, b) \in C \mid b \in F\}| = 1$,
- $\forall a \in A \cup S \cup E_e : |\{(b, a) \in C \mid b \in F\}| = 1$,
- $\forall a \in J : |\{(b, a) \in C \mid b \in F\}| \geq 2$,
- $\forall a \in S : |\{(a, b) \in C \mid b \in F\}| \geq 2$, *and*
- *all elements $e \in F$ in the graph (F, C) are on a path from a start event $a \in E_s$ to an end event $b \in E_e$.*

For a block-structured BP model it is furthermore required that the process is hierarchically organised [15], i.e. it consists of unique join-split-pairs, each representing either a single entry or a single exit point of a non-sequential BP construct, e.g. Choice, Parallel, Loop, etc. The example process in Fig. 1 is a block-structured process. A similar representation gaining popularity in recent years is the *process tree*, as defined based on Petri nets/workflow nets in [9].

When a business process is automatically or semi-automatically executed with a BP execution engine, e.g. with a Business Process Management System (BPMS), an *event log* is produced, i.e. a all occurred events are logged and

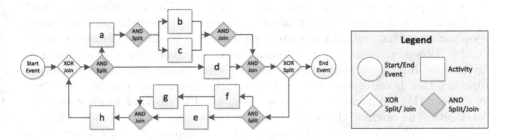

Fig. 1. Example business process with all element types included

stored. These logs and their contained events may capture different aspects of a process execution, e.g. a different granularity of events are logged. In this paper however, we only focus on a minimal set of event features: In order to allow the discovery of the control-flow, every event is required to have a reference (1) to the associated *process instance* and (2) to the corresponding *activity*. Furthermore, we assume that the log contains exactly one event for each activity execution, i.e. activity lifecycle events are not regarded. All events resulting from the execution of the same *process instance* are captured in one *trace*. A trace is assumed to be independent from other traces, i.e. the execution order of a process instance is not in any way dependent on the execution of a second instance. Accordingly, an event e is represented by a pair $e = (t, a)$ where $t \in \mathbf{N}$ is the unique identifier of the trace and $a \in A$ is a unique reference to the executed activity.

The research area of *Process Discovery* is concerned with the extraction of a business process model from event logs without using any a-priori information [23]. Conventional challenges in process discovery originate from the motivation to achieve a high quality of results, i.e. discovered processes should support as accurately as possible the behaviour contained in the log. In particular that means, process discovery algorithms have to deal with multiple objectives, e.g. precision, simplicity, fitness - over-fitting vs. under-fitting (see [23]). Process discovery algorithms are usually assumed to be carried out in an static way as an "offline" method. This is reflected by the fact that the input for these algorithms is an entire log as conceptually shown by the following definition:

Definition 2. *Let the log $L_n = [e_0, e_1, ...e_n]$ be a sequence of $n+1$ events ordered by time of occurrence ($\forall i < j \wedge e_i, e_j \in L_n : time(e_i) \leq time(e_j)$) and BP_n be the business process model representing the behaviour in L_n, then process discovery is defined as a function that maps a log L_n to a process BP_n:*

$$ProcessDiscovery : [e_0, e_1, ..., e_n] \Rightarrow BP_n$$

3   Related Work

A large number of process discovery algorithms exist, e.g. Inductive Miner [9], HeuristicsMiner [25], alpha-miner [20] and CCM [15]. These and many algorithms have in common that at first a *footprint* of the log is created based on which the process is constructed. Similar to the CCM, the following related algorithms also discover block-structured processes: (1) Genetic process discovery algorithms that restrict the search space to block-structured process models, e.g. [4]. However, these are non-deterministic and generally have a high execution time due to exponentially expanding search space. (2) Another relevant approach that is conceptually similar to the CCM is proposed in [9], the Inductive Miner (IM): A top-down approach is applied to discover block-structured Petri nets. The original algorithm evaluates constraints based on local relationships between activities in order to identify the representing construct in an inductive fashion. In recent work, the IM has also been extended to deal with noise [10]. Generally, in all discovery approaches based on footprints known to the authors the footprint is represented by a *direct neighbours* matrix representing information about the local relations between the activities, e.g. for the BP of Fig. 1: h can only appear *directly* after g or e. As discussed in Sect. 4 the CCM on the other hand extracts the process from a footprint based on global relations between activities, e.g. h appears *at some point* after g or e.

However, of little importance for conventional process discovery algorithms is their practicality with regards to an application during run-time: as defined in Definition 2 process discovery is a static method that analyses an event log in its entirety. An alternative to this approach is the immediate processing of events when they occur to information of an higher abstraction level in order to enable a real-time analysis. This approach is called Complex Event Processing (CEP): a method that deals with the event-driven behaviour of large, distributed enterprise systems [11]. More specifically, in CEP events produced by the systems are captured, filtered, aggregated, and finally abstracted to complex events representing high-level information about the situational status of the system, e.g. performance, control-flow, etc. The need for monitoring aspects of business processes at run-time by applying CEP methodologies has been identified by Ammon et al., thus coining the term Event-Driven Business Process Management (EDBPM) - a combination of two disciplines: Business Process Management (BPM) and Complex Event Processing [1]. The dynamic process discovery solution proposed in this paper is an application of EDBPM (see Sect. 5). The motivation is to have a run-time reflection of the employed processes based on up-to-date rather than historical information which essentially allows business analysts to react quicker to changes or occurring bottlenecks etc. in order to optimise the overall performance of the monitored processes. In accordance to this objective process discovery algorithms for event streams have to deal with two additional challenges as opposed to the traditional process discovery algorithms:

1. The application of process discovery on event streams is executed in a real-time setting and thus is required to conform to special memory and execution time constraints. Especially with regards to many modern systems producing "big data", i.e. data that is too large and complex to store and process all of it [12]. This means in particular, that online algorithms should be able to (1) process an infinite number of events without exceeding a certain memory threshold and (2) process each event within a small and near-constant amount of time [5].
2. Real-life BPs are often subject to externally or internally initiated change which has to be reflected in the results of an process discovery algorithm analysing an event stream. The observed characteristic of dynamically changing processes is also called *concept drift* and has been identified as a major challenge for process discovery algorithms [2, 22, 23]. Generally, discovery algorithms on event streams should be able to (1) reflect newly appearing behaviour as well as (2) forget outdated behaviour.

Although not specifically, *incremental process mining* as introduced in [3] does attempt to anticipate the problem of concept drift to some extent. Here, the assumption is that a log does not yet contain the entire behaviour of the process (i.e. an incomplete log) at the time of the discovery of the initial (declarative) process. Additional behaviour, that occurred after the initial discovery and captured in a second log, is analysed separately and the new information is then added to the existing BP model. This is possible due to structure of declarative BP specifications. The process of incrementally analysing log segments and then extending the BP model accordingly, i.e. incremental process mining, is motivated by the assumption that the update of an already existing (declarative) BP model is easier than to always analyse the complete log from scratch [3]. Another approach called *incremental worklfow mining* is based on the same principle but does discover and adapt a Petri Net from incrementally processing log segments [6,7]. It is a semi-automatic (and prototypical) approach specifically designed for dealing with process flexibility in *Document Management Systems* that does not anticipate incomplete or noisy logs. A third incremental approach is presented in [18] which utilises the theory of regions to create transition systems for successive sub-logs and eventually transform them into a Petri Net. Albeit based on a slightly different concept, incremental process mining approaches can be considered for process discovery on event streams since the event processing could be designed to group a number of successive traces into sub-logs which are then individually analysed and incrementally extend the overall BP. However, a conceptual weakness of incremental mining approaches is the lacking ability of forgetting outdated behaviour.

In the context of process discovery on event streams, a synonymous term sometimes used is *Streaming Process Discovery* (SPD). SPD was coined by Burattin et al. in [5]. In their work the HeuristicsMiner [25] has been modified for this purpose and a comprehensive evaluation of different event stream processing types was carried out. The fundamentals of the HeuristicsMiner remain the same but the direct-neighbours-footprint is dynamically adapted or rebuilt while processing the individual events. From this the Causal Net is periodically extracted, e.g. for every event or every 1000 events, using the traditional HeuricticsMiner. For instance, for the evaluation of the different streaming methods the HM discovery was triggered every 50 events [5]. Three different groups of event streaming methods have been implemented and investigated:

Event Queue: The basic methodology of this approach is to collect events in a queue which is representing a log that can be analysed in the traditional way of process discovery. Three basic types can be differentiated: (1) In the *sliding window* approach the queue is a FIFO (First-In-First-Out), i.e. when the maximum queue length (queue memory) is reached for every new event inserted, the oldest event in the queue is removed; (2) In the *periodic reset* approach the queue is reset whenever the maximum queue length is reached. The main advantage of these approaches are that the event queue can be regarded as event log and enables the analysis via traditional discovery/mining algorithms on event logs. Two of the main disadvantages are: Each event is handled at least twice: once to store it in the queue and once or more to discover the model from the queue; Also, it does only allow for a rather simplistic interpretation of "history", i.e. an older event is either still in the queue and has same influence as a newer event, or it is completely forgotten.

Stream-Specific Approaches: Stream-specific approaches already process events into footprint information, i.e. queues that consist of a fixed size hold information about the latest occurring activities and directly-follows relations. When a new event occurs all values in the queues are updated and/or replaced. Burattin et al. distinguish between the following three update operations: (1) *Stationary*, i.e. the queues function as a "sliding window" over the event stream and every queue entry has the same weight, (2) *Ageing*, i.e. the weight of the latest entry is increased and the weights of older entries in the queue are decreased, and (3) *Self-Adaptive Ageing*, i.e. the factor with which the influence of older entries decreases is dependent on the fitness of the discovered model in relation to latest events stored in an additional sample queue of a fixed size: quickly decreasing for a low fitness and slowly decreasing for a high fitness. Generally, stream-specific approaches are assumed to be a more balanced approach since events are only handled once and directly processed into footprint information [5]. Burattin et al. also argue that ageing-based approaches have a more realistic interpretation of "history" since older events have less influence than newer events [5]. One disadvantage is that the footprint is captured through a set of queues with a fixed size: if this size is set too low, behaviour is prematurely forgotten; if this size is set too high some of the old behaviour is never forgotten.

Lossy Counting: Lossy Counting is a technique adopted and modified from [13] that uses approximate frequency count and divides the stream into a fixed number of buckets.

Another approach for discovering concept drifts on event streams of less relevance to the paper's topic is presented in [12]: A discovery approach for declarative process models using the sliding window approach and lossy counting to update a set of valid business constraints according to the events occurring in the stream.

4   Static Constructs Competition Miner

The CCM as described in [15] is a deterministic process discovery algorithm that operates in a static fashion and follows a divide-and-conquer approach which, from a given event log, directly mines a block-structured process model that represents the main behaviour of the process. The CCM has the following main features [15]: (1) A deadlock-free, block-structured business process without duplicated activities is mined; (2) The following BP constructs are supported and can be discovered for single activities: Normal, Optional, Loopover, and Loopback; or for a set of activities: Choice, Sequence, Parallel, Loop, Loopover-Sequence, Loopover-Choice, Loopover-Parallel (see Fig. 2), and additionally all of them as optional constructs - these are constructs supported by the majority of business process

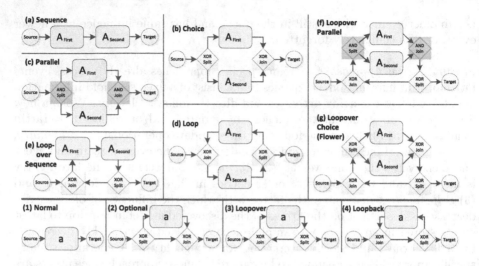

Fig. 2. Business process constructs supported by the CCM [15]

Algorithm 1. Methodology of the CCM in Pseudocode

Data: *Log L*
Result: *BP bp*

```
1  begin
2  │  A ← getSetOfAllActivitiesInLog(L);
3  │  BP bp ← buildInitialBPWithStartAndEnd();
4  │  bp ← getFootprintAndBuildConstruct(A, L, bp);
5  └  return bp;

6  Function getFootprintAndBuildConstruct(Am, Log L, BP bp)
7  │  Footprint fp = extractFootprintForActivities(Am, L);
8  │  if |Am| = 1 then
9  │  │  Construct c ← analyseConstructForSingleActivity(fp);
10 │  │  bp ← createSingleActivityConstruct(c, Am);
11 │  else
12 │  │  ConstructsSuitability[] cs ← calculateSuitabilityForConstructs(fp, Am);
13 │  │  (Construct c, Afirst, Asecond) ← constructCompetition(cs, Am);
14 │  │  bp ← createBlockConstruct(c, bp);
15 │  │  bp ← getFootprintAndBuildConstruct(Afirst, L, bp);
16 │  │  bp ← getFootprintAndBuildConstruct(Asecond, L, bp);
17 └  return bp;
```

---

standards like BPMN or YAWL; (3) If conflicting or exceptional behaviour exists in the log, the CCM picks the "best" fitting BP construct.

Algorithm 1 shows the conceptual methodology of the CCM algorithm in pseudocode. The CCM applies the divide-and-conquer paradigm and is implemented in a recursive fashion (see lines 7, 16, and 17). At the beginning getFootprintAndBuildConstruct is initially called for all involved activities ($A_m = A$) with the process bp consisting of only a start and end element. The recursive function is first creating a footprint fp from the given log L only considering the activities specified in set A_m (at the beginning all involved activities). In a next step it will be decided which is the best construct to represent the behaviour captured by fp: (1) if the activity set A_m only consists of one element, it will be decided which of the single activity constructs (see bottom of Fig. 2) fits best - the process bp will then be enriched with the new single activity construct (see line 11); (2) If the activity set A_m contains more than one element, the suitability for each of the different constructs is calculated for any two activities $x, y \in A_m$ based on "soft" constraints and behaviour approximations, e.g. activities a and b are in a strong Sequence relationship. The result of this calculation (line 13) is a number of suitability matrices, one for each construct. In the subsequent competition algorithm it is determined what is the best combination of (A) the construct type $c \in \{Sequence, Choice, Loop, ...\}$, and (B) the two subsets A_{first} and A_{second} of A_m with $A_{first} \cup A_{second} = A_m$, $A_{first} \cap A_{second} = \{\}$, and $A_{first}, A_{second} \neq \{\}$, that best accommodate all x, y-pair relations of the corresponding matrix of construct c (line 14). The construct is then created and added to the existing process model bp (line 15), e.g. XOR-split and -join if the winning construct c was $Choice$. At this stage the recursive method calls will be executed to analyse and construct the respective behaviour for the subsets A_{first} and A_{second}. The split up of the set A_m continues in a recursive fashion until it cannot be divided any more, i.e. the set consists of a single activity (see case (1)). The process is completely constructed when the top recursive call returns.

Of particular interest for the transformation of the CCM algorithm to a solution for dynamic process discovery is the composition of the footprint and its calculation from the log. As opposed to many other process discovery algorithms, e.g. alpha-miner [20], the footprint does not consist of absolute relations, e.g. h is followed by a (see example in Fig. 1), but instead holds relative relation values, e.g. a is eventually followed by g in 0.4 $\cong$ 40 % of the traces. Furthermore, the footprint only contains global relations between activities in order to guarantee a low polynomial execution time for the footprint interpretation [15]. The footprint of the CCM contains information about: (1) the occurrence of each involved activities $x \in A_m$, i.e. how many times x appears at least once per trace, how many times an x appears on average per trace, and how many times the trace started with x; (2) the global relations of each activity pair $x, y \in A_m$, i.e. in how many traces x appears sometime before the $first$ occurrence of y in the trace, and in how many traces x appears sometime before any occurrence of y in the trace[2]. All measures in the footprint are relative to the number of traces in the log. Furthermore, not only one overall footprint is created for the CCM but also for every subset A_{first} and A_{second}, that is created during execution, a new sub-footprint is created (see Algorithm 1).

---

[2] This stands in contrast to existing discovery solutions since in the CCM the footprint and its interpretation is not based on *local* relationships between activity occurrences, e.g. direct neighbours, but based on *global* relationships between them.

5   Dynamic Constructs Competition Miner

As established in Sect. 1, increasingly dynamic processes and the need for immediate insight require current research in the domain of process mining to be
driven by a set of additional challenges. To address these challenges the concept
of Scalable Dynamic Process Discovery (SDPD), an interdisciplinary concept
employing principles of CEP, Process Discovery, and EDBPM, has been introduced in [16]: "SDPD describes the method of monitoring one or more BPMSs
in order to provide at any point in time a reasonably accurate representation of
the current state of the processes deployed in the systems with regards to their
control-flow, resource, and performance perspectives as well as the state of still
open traces." That means, any potential changes in the mentioned aspects of
the processes in the system that occur during run-time have to be recognized
and reflected in the continuously updated "current state" of the process. Due to
its purpose, for solutions of SDPD an additional set of requirements applies. For
this paper, the most relevant of them are [16]:

- *Detection of Change*: An SDPD solution is required to detect change in two
  different levels defined in [17]: (1) Reflectivity: A change in a process instance
  (trace), i.e. every single event represents a change in the state of the associated
  trace. (2) Dynamism: A change on the business process level, e.g. because
  events/traces occurred that contradicts with the currently assumed process.
- *Algorithmic Run-Time*: An SDPD solution is applied as CEP concept and has
  to be able deal with large business processes operating with a high frequency,
  i.e. the actual run-time of the algorithms becomes very important. The key
  algorithms should be run-time effective to cope with increasing workload at
  minimal possible additional computational cost.

Motivated by these challenges the initial process discovery approach was altered
to allow for dynamic process discovery. As opposed to the traditional *static*
methodology (see Definition 2), *dynamic* process discovery is an iterative approach as defined in the following:

Definition 3. *Let log $L_n = [e_0, e_1, ...e_n]$ be a sequence of $n+1$ events ordered by
time of occurrence ($\forall i < j \wedge e_i, e_j \in L_n : time(e_i) \leq time(e_j)$) and BP_n be the
business process model representing the behaviour in L_n, then dynamic process
discovery is defined as a function that projects the tuple (e_n, BP_{n-1}) to BP_n:*

$$DynamicProcessDiscovery : (e_n, BP_{n-1}) \Rightarrow BP_n$$

As described in Sect. 4, the CCM is a static mining algorithm and has to
be modified in order to enable SDPD. The result of this modifications is called
Dynamic CCM (DCCM). However, two restrictions for the DCCM with regards
to the previously mentioned requirements of SDPD apply: (1) instead of discovering change on the BP perspectives control-flow, resources, and performance perspective, the DCCM described in this paper only focuses on discovering change
in the control-flow, and (2) only change on the abstraction level of Dynamism

is detected, i.e. whether or not the control-flow of the process has changed - the detection of change on the abstraction level of Reflectivity will not be supported by the DCCM. Additionally to the requirements of SDPD the DCCM features the following important aspects: (1) *robust*: if conflicting, exceptional, or not representable behaviour occurs in the event stream, the DCCM does not fail but always picks the BP construct that best accommodates the recorded behaviour; (2) *deterministic*: the DCCM yields the exact same output BP for the same input stream of events.

Four different modifications were applied to the default CCM to create the DCCM. These modifications are summarised in the following list and described in more detail in the following sub-sections:

1. Splitting up the algorithm in two separate parts: one for dynamically updating the current footprint(s) complying to the requirement of extremely low algorithmic run-time, and one for interpreting the footprint into a BP model which has less restrictions with regards to its run-time.
2. In the CCM the footprint is calculated in relation to all occurring traces. This is not applicable for SDPD since the number of traces should not have an influence on the execution-time of any component of an SDPD solution. For this reason the footprint has to be calculated in a dynamic fashion, i.e. an event-wise footprint update independent from the previously occurred number of events or traces.
3. The original behaviour of the CCM to carry out a footprint calculation for every subset that has been created by the divide-and-conquer approach is not optimal as then the DCCM would have to extract up to $2 * n + 1$ different footprints if only one activity was split-up from the main set for each recursion.[3] This has been improved for the DCCM: for the most common constructs Choice and Sequence the sub-footprints are automatically derived from the parent footprint.
4. In rare cases it can happen that for every appearing event the state of the process is alternating between a number of different control-flows. This is caused by "footprint equivalent" BP models, i.e. two models are footprint equivalent if they both express the behaviour captured by the footprint. We introduce a measure which favours the last control-flow state in order to prevent the described behaviour.

5.1   Methodology of the Dynamic CCM

The original CCM algorithm had to be split up into two separate parts in order to comply to the SDPD's requirement of low algorithmic run-time for the event processing. A component triggered by the occurrence of a new event to update

---

[3] e.g. for $A = \{a, b, c, d\}$: $(a, b, c, d) \rightarrow ((a, b, c), (d)) \rightarrow (((a), (b, c)), (d)) \rightarrow ((((a), ((b), (c))), (d))$, seven different footprints for sets $\{a, b, c, d\}$, $\{a, b, c\}$, $\{b, c\}$, $\{a\}$, $\{b\}$, $\{c\}$, $\{d\}$ need to be created - $(,)$ denote the nested blocks that emerge while splitting the sets recursively.

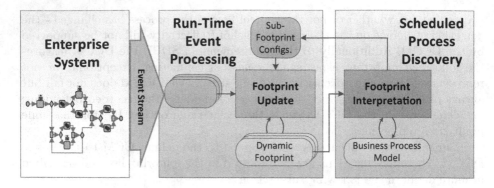

Fig. 3. Conceptual methodology of the dynamic CCM

the dynamic footprint and a component decoupled from the event processing which interprets the footprint into a BP Model. The conceptual methodology of the DCCM is depicted in Fig. 3. The components, models, and functionality of the DCCM are described in the following: Events from the monitored *Enterprise System*, in which the end-to-end process is deployed, are fed into an event stream. The *Footprint Update* component is the receiver of these events and processes them directly into changes on the overall *Dynamic Footprint* which represents the abstract state of the monitored business process. If additional footprints for subsets of activities are required as specified by the *Sub-Footprint Configurations*, e.g. if a Loop or Parallel construct was identified, then these sub-footprints are also updated (or created if they were not existent before). The *Dynamic Footprint(s)* can then at any point in time be compiled to a human-centric representation of the business process by the *Footprint Interpretation* component, i.e. the abstract footprint representation is interpreted into knowledge conforming to a block-structured BP model. In the DCCM this interpretation is scheduled dependent on how many new completed traces appeared, e.g. the footprint interpretation is executed once every 10 terminated traces. If the *interpretation frequency* $m \in \mathbb{N}$ of the DCCM is set to 1 a footprint interpretation is executed for every single trace that terminated. The *Footprint Interpretation* algorithm works similar to the CCM algorithm shown in Algorithm 1; but instead of extracting footprints from a log (line 8), the modified algorithm requests the readily available *Dynamic Footprint(s)*. If a sub-footprint is not yet available (e.g. at the beginning or if the process changed) the *Footprint Interpretation* specifies the request for a sub-footprint in the *Sub-Footprint Configurations* in the fashion of a feedback loop. Thus, *Sub-Footprint Configurations* and *Dynamic Footprints* act as interfaces between the two components, *Footprint Update* and *Footprint Interpretation*. The *Footprint Interpretation* cannot continue to analyse the subsets if no sub-footprint for these exist yet. In this case, usually occurring in the warm-up or transition phase, an intermediate BP model is created with activities containing all elements of the unresolved sets as depicted in Fig. 4.

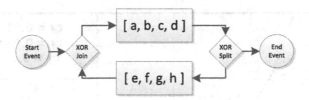

Fig. 4. Result of the *Footprint Interpretation* on an event stream produced by the example from Fig. 1 if no sub-footprints for $\{a, b, c, d\}$ and $\{e, f, g, h\}$ are available yet - only the top-level loop has been discovered

5.2   Run-Time Update of the Dynamic Footprint

The *Footprint Update* component processes events to changes in the *Dynamic Footprint*, i.e. updates the abstract representation of the process state. The original footprint extraction of the CCM algorithm calculates all values in relation to the number of occurred traces, i.e. every trace's influence on the footprint is equal: $\frac{1}{|traces|}$. To keep the algorithmic run-time to a minimum and allow for scalability the footprint update calculation should only take a fixed amount of time, independent from the total number of previously occurred events or traces. An increase of the total number of involved activities can cause, however, a linear increase of the execution-time due to the recalculation of the relations between the occurred activity and, in the worst case, all other activities. The independence from previous traces is the reason the footprint is calculated in a dynamic fashion, i.e. the dynamic footprint is incrementally updated in a way that older events "age" and thus have less influence than more recent events.

The general ageing approach that is utilized in the *Footprint Update* of the DCCM is based on the calculation of an individual *trace footprint*[4] (TFP) for each trace which influences the dynamic overall footprint (DFP). For the n-th new TFP_n the DFP is updated in the following way: Given a specified *trace influence factor* $t_{if} \in \mathbb{R}$ with $0 < t_{if} \le 1$ the old DFP_{n-1} is aged by the *ageing factor* $a_f = 1 - t_{if}$, i.e.

$$DFP_n = t_{if} * TFP_n + (1 - t_{if}) * DFP_{n-1} \tag{1}$$

E.g., for trace influence factor $t_{if} = 0.01$: $DFP_n = 0.01 * TFP_n + 0.99 * DFP_{n-1}$. Two different ageing methods have been developed which will be evaluated against each other in Sect. 6: *Occurrence-based Ageing* and *Time-based Ageing*.

Occurrence-Based Ageing is an ageing method similar to the approach of Burretin et al. [5] discussed in Sect. 3. In this case the trace influence t_{if} is a fixed value and the DFP ages the same proportion for every time a trace footprint

---

[4] the occurrence values for activities as well as the global relations (see end of Sect. 4) are represented in the trace footprint by absolute statements $true \equiv 1$ if it occurred and $false \equiv 0$ if not.

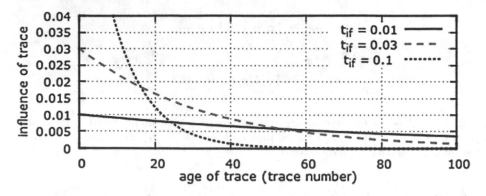

Fig. 5. Development of the influence of a trace for different trace influence factors(t_{if})

is added indeterminate from how much time has passed since the last footprint update. For example, assuming $t_{if} = 0.01$ the trace footprint TFP_n has the influence of 0.01 when it first occurs (see Eq. 1); after another TFP_{n+1} has occurred the influence of TFP_n decreases to $0.01 * 0.99$, and after another $0.01 * 0.99^2$ and so on. By applying this incremental method, older TFP are losing influence in the overall dynamic footprint. Figure 5 shows how the influence of a trace is dependent on its "age": If $t_{if} = 0.1$, the influence of a trace that appeared 60 traces ago became almost irrelevant. At the same time if $t_{if} = 0.01$ the influence of a trace of the same age is still a little more than half of its initial influence when it first appeared. Essentially, the purpose of the *trace influence factor* t_{if} is to configure both, the "memory" and the adaptation rate, of the footprint update component, i.e. a high t_{if} means quick adaptation but short memory but a small t_{if} means a slow adaptation but a long memory. Finding the correct trace influence is an issue of balancing these two inversely proportional effects, e.g. it might be generally desirable to have a high adaptation rate ($t_{if} = 0.1$) but if some behaviour of the process only occurs once in every 60 traces it will already be "forgotten" when it reappears (see Fig. 5) essentially resulting in a continuously alternating business process. However, while applying this method it was observed that at the beginning of the event streaming an unnecessarily long time to "warm-up" was required until the DFP reflected the correct behaviour of the business process. In order to shorten the "warm-up" phase of the *Footprint Update* a more dynamic method was adopted: If the overall amount of so far occurred traces $|traces| < \frac{1}{t_{if}}$ then the influence of the dynamic overall footprint is $\frac{|traces|}{|traces|+1}$ and of the new trace footprint $1 - \frac{1}{|traces|+1}$. As a result all traces that occur while $|traces| < \frac{1}{t_{if}}$ have the same influence in the DFP of $\frac{1}{|traces|}$. For instance if $t_{if} = 0.01$ and $|traces| = 9$ then a new dynamic footprint is calculated with $DFP_{10} = \frac{1}{10} * TFP + \frac{9}{10} * DFP_9$ and for the next trace $DFP_{11} = \frac{1}{11} * TFP + \frac{10}{11} * DFP_{10}$. As soon as $|traces| \geq \frac{1}{t_{if}}$ the standard occurrence ageing with a fixed influence factor is adopted:

$$DFP_n = \begin{cases} TFP_n & \text{if } n = 0 \\ \frac{1}{n} * TFP_n + \frac{n-1}{n} * DFP_{n-1} & \text{if } 0 < n < \frac{1}{t_{if}} \\ t_{if} * TFP_n + (1 - t_{if}) * DFP_{n-1} & \text{if } n \geq \frac{1}{t_{if}} \end{cases} \quad (2)$$

Because of this implementation the "warm-up" phase of the *Footprint Update* could be drastically reduced, i.e. processes were already completely discovered a few traces after the start of the monitoring which will be shown in Sect. 6.

Time-Based Ageing is an ageing method based on the time that has passed since the last trace occurred. The more time has passed the less influence the old DFP_{n-1} has on the updated DFP_n. This is achieved in a similar way than in the occurrence-based ageing but instead of having an ageing factor a_f relative to the trace occurrence it is now relative to the time passed. That means that ageing factor a_f and trace influence factor t_{if} are not fixed for each trace occurrence but calculated based on an *ageing rate* a_r per passed *time unit* t_{ur}, i.e. in particular time t_n has passed since the last trace occurred then

$$a_f = a_r^{\frac{t_n}{t_{ur}}} \text{ and } t_{if} = 1 - a_f$$

If $t_n = t_{ur}$ then the dynamic overall footprint ages exactly the same as with the occurrence-based ageing, if $t_n > t_{ur}$ then it ages quicker, and if $t_n < t_{ur}$ it ages slower. For instance, with ageing rate $a_r = 0.99$ and time unit $t_{ur} = 1s$: If the new trace occurred $t_n = 2s$ after the last footprint update then the new dynamic overall footprint $DFP_n = (1 - 0.99^2) * TFP_n + 0.99^2 * DFP_{n-1}$. Through time-based ageing the influence development of passed traces behaves similarly to the occurrence-based ageing in Fig. 5 apart from that the ageing is not based on trace-oldness but on time-oldness (the numbers on the x-axis now represent the passed time units since the trace occurred). For the time-based ageing a similar problem was observed during the warm-up phase than with the occurrence-based ageing: Since the DFP only consists of zeros when initialised it takes an unnecessary long time to converge towards a footprint representing the correct behaviour of the business process. For this reason an alternative linear ageing relative to the overall passed time since the first trace recorded t_{ur} was adopted as well. The final ageing factor a_f is the minimum of both calculated values as shown in Eq. 3.

$$a_f = \min(1 - \frac{t_n}{t_{all}}, a_r^{\frac{t_n}{t_{ur}}}) \quad (3)$$

$$DFP_n = (1 - a_f) * TFP_n + a_f * DFP_{n-1} \quad (4)$$

Considering the example from earlier where $a_r = 0.99$, time unit $t_{ur} = 1s$, the new trace TFP_n occurred $2s$ after the last update, and the first trace recorded was $t_{all} = 4s$ then $a_f = \min(1 - \frac{2s}{4s}, 0.99^2) = 0.5$ and according to Eq. 4: $DFP_n = (1 - 0.5) * TFP_n + 0.5 * DFP_{n-1}$. In this way the warm-up phase can be shortened similar to the occurrence-based approach but still be based on time.

Another important dynamism feature that had to be implemented was the possibility to add an activity that has not appeared before. A new activity is first recorded in the respective trace footprint. When the trace is terminated it will be added to the overall footprint in which it is not contained yet. The factored summation of both footprints to build the new dynamic footprint is carried out by assuming that a not previously in the dynamic overall footprint contained relation value is 0. Furthermore, activities that do not appear any more during operation should be removed from the dynamic footprint. This was implemented in the DCCM in the following way: If the *occurrence once* value of an activity drops below a removal threshold $t_r \in \mathbb{R}, t_r < t_{if}$ it will be removed from the dynamic footprint, i.e. all values and relations to other activities are discarded.

The fact that especially many Choice and Sequence constructs are present in common business processes, motivates an automated sub-footprint creation in the *Footprint Interpretation* based on the parent footprint rather then creating the sub-footprint from the event stream. This step helps to decrease the execution-time of the *Footprint Update* and was achieved by introducing an extra relation to the footprint[5] - the *direct neighbours* relation as used by other mining algorithms (see Sect. 3). In the *Footprint Interpretation* this relation is then used for creating the respective sub-footprints for Sequence and Choice constructs but not for identifying BP constructs since the *direct neighbours* relation does not represent a global relation between activities.

5.3   Modifications in the Footprint Interpretation Component

As analysed in the beginning of this section, the original behaviour of the CCM to retrieve a sub-footprint for each subset that has been created by the divide-and-conquer approach is not optimal. This is why, in the *Footprint Interpretation* the DCCM calculates the sub-footprints for the most common constructs, Choice and Sequence, from the available parent footprint: (1) For the Choice construct the probability of the exclusive paths are calculated with $p_{first} = \sum_{x \in A_{first}} Fel(x)$ and $p_{second} = \sum_{x \in A_{second}} Fel(x)$ with $Fel(x)$ being the occurrences of x as first element (see CCM footprint description in Sect. 4). Then the relevant values of the parent footprint are copied into their respective new sub-footprints and normalized, i.e. multiplied with $\frac{1}{p_{first}}$ and $\frac{1}{p_{second}}$, respectively. (2) The sub-footprints for the Sequence construct are similarly built, but without the normalization. Instead, the direct neighbours relation, now also part of the dynamic footprint, is used to calculate the new overall probabilities of the sub-footprints.

If two or more BP constructs are almost identically suitable for one and the same footprint, a slight change of the dynamic footprint might result in a differently discovered BP. This may cause an alternating behaviour for the footprint

---

[5] In rare cases (if Loop and Parallel constructs dominate) this modification can have a negative effect on the execution-time since extra information needs to be extracted without the benefit of mining less sub-footprints.

interpretation, i.e. with almost every footprint update the result of the interpretation changes. This is undesirable behaviour which is why the competition algorithm was additionally modified as follows: All combinations of BP construct and subsets are by default penalized by a very small value, e.g. $\frac{t_{if}}{10}$, with the exception of the combination corresponding to the previously discovered BP model, hence reducing the risk of discovering alternating BP models.

6 Evaluation

The static CCM algorithm has been tested in detail for its accuracy in [15]: (1) in a qualitative analysis the CCM was able to rediscover 64 out of 67 processes for which a log was produced through simulation. (2) in the second part of the evaluation the discovery performance of the CCM was compared to the mining algorithms HeuristicsMiner (HM) [25], Inductive Miner (IM) [10], and the Flower Miner (FM), all of which are readily available in the ProM nightly build [21]. For ten given logs (including real-life logs and publicly available logs) the results of the algorithms (each configured with their default parameters) were evaluated for their *trace fitness* f_{tf}, *precision* f_{pr}, *generalization* f_g, and *simplicity* f_s with the help of the *PNetReplayer* plugin [24]. The averaged results of the detailed analysis are shown in Table 1 [15]; Note, that a lower simplicity value is better.

Table 1. Conformance results of the different discovery algorithms [15]

Trace Fitness f_{tf}				Precision f_{pr}				Generalization f_g				Simplicity f_s			
HM	IM	FM	CCM	HM	IM	FM	CCM	HM	IM	FM	CCM	HM	IM	FM	CCM
0.919	0.966	1.0	0.979	0.718	0.622	0.124	0.663	0.941	0.915	0.992	0.930	155.3	122.8	56.4	111.9

In the remainder of this section evaluation results of the DCCM are presented with regards to its capability of initially discovering the correct process and how it reacts to certain changes of a real-time monitored business process. Furthermore, a comparative analysis is carried out to determine the advantages and disadvantages of the two ageing strategies, occurrence-based and time-based ageing.

6.1 Experiment Setup

The experiments revolve around the concept of process rediscovery, i.e. that a source business process is executed and produces an event stream which is fed to the DCCM which then should discover a process behaviourally equivalent to the executed source process. Figure 6 shows three measures (t_w, t_d, and t_{tr}) which we use to evaluate the quality of the DCCM. In the figure BP_1 and BP_2 are the business processes deployed in the monitored system and BP_1' to BP_n' are the models discovered by the DCCM. Additionally, BP_1 and BP_m' are equivalent ($BP_1 \equiv BP_m'$) as well as BP_2 and BP_n' ($BP_2 \equiv BP_n'$). For this part of the evaluation the following measures are of interest:

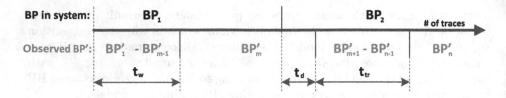

Fig. 6. Measures for detection of BP change in system

- Warm-up: $t_w \in \mathbb{N}$ the amount of completed traces the DCCM needs as input at the start until the resulting model equivalently represents the process in the system, i.e. until $BP_1 \equiv BP'_m$.
- Change Detection: $t_d \in \mathbb{N}$ the amount of completed traces it takes to detect a certain change in the monitored process - from the point at which the process changed in the system to the point at which a different process was detected. When the change is detected the newly discovered process is usually not equivalent to the new process in the system BP_2 but instead represents parts of the behaviour of both processes, BP_1 and BP_2.
- Change Transition Period: $t_{tr} \in \mathbb{N}$ the amount of completed traces it takes to re-detect a changed process - from the point at which the process change was detected to the point at which the correct process representation was identified, i.e. until $BP_2 \equiv BP'_n$. In this period multiple different business processes may be detected, each best representing the dynamic footprint at the respective point in time.

The basis of our evaluation is the example model in Fig. 1 which is simulated and the resulting event stream fed into the DCCM. In order to get reliable values for the three measures t_w, t_d, and t_{tr}, this is repeated 60 times for each configuration. From these 60 runs the highest and lowest 5 values for each measure are discarded and the average is calculated over the remaining 50 values. For each experiment the CCM core is configured with its default parameters (see [15]).

6.2   Warm-up Evaluation

The first experiment will evaluate how the DCCM behaves at the beginning when first exposed to the event stream, more particularly, we want to determine the duration of the warm-up phase t_w. Figure 7 shows the development of the first few BP models extracted by the DCCM using *Occurrence Ageing* with *trace influence factor* $t_{if} = 0.01$ (see Sect. 5.2) and *interpretation frequency* $m = 10$, i.e. an interpretation is executed every 10 completed traces: After the first trace the discovered process is a sequence reflecting the single trace that defines the process at that point in time. At trace 10, which is the next scheduled footprint interpretation, the algorithm discovers a Loop construct but cannot further analyse the subsets since the corresponding sub-footprint was not requested yet. Because of that, the feedback mechanism via the *Sub-Footprint Configurations* is utilized by the *Footprint Interpretation* algorithm to register the creation of

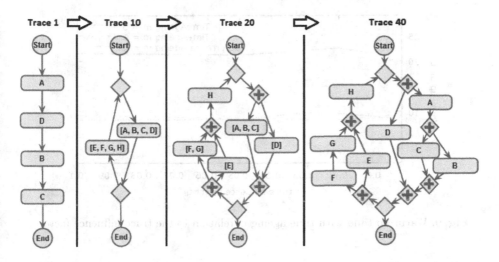

Fig. 7. The evolution of the discovered BP model during the warm-up phase

the missing sub-footprints. In the next scheduled run of the footprint interpretation, the Parallel construct of a, b, c, and d is discovered but again the analysis can not advance since a sub-footprint for the individual activity subsets has not been created yet. Activities e, f, g, and h seem to have appeared only in exactly this sequence until trace 20. Skipping one of the interpretation steps, we can see that at trace 40 the complete process has been mined, i.e. $t_w = 40$.

In Fig. 8 the development of warm-up duration t_w for of the DCCM with *Occurrence Ageing* for different $m \in \{1, 2, 3, 6, 10\}$ and $t_{if} \in \{0.001, 0.002, 0.005, 0.01, 0.018, 0.03\}$ is depicted. The warm-up phase seems generally very short and not strongly influenced by t_{if}. For $m = 10$ the warm-up phase cannot be any shorter because the example process consists of a block-depth of 3:

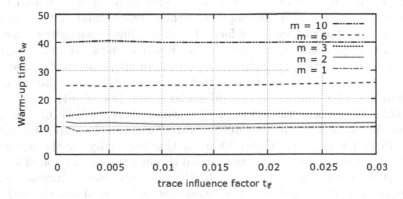

Fig. 8. Warm-up time with the occurrence ageing in relation to the trace influence factor

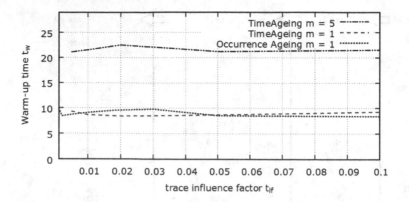

Fig. 9. Warm-up time with time ageing in relation to the trace influence factor

Parallel-in-Parallel-in-Loop, i.e. 3 subsequent requests for sub-footprints have to be made. This is an indicator that the modification effort to shorten the warm-up phase had a positive effect. No significant changes of t_w can be noticed when increasing or decreasing the *trace influence factor* t_{if}. This is caused by the optimisation rule that essentially nullifies the default ageing via t_{if} (see Eq. 2 in Sect. 5.2).

A similar sensitivity experiment was carried out for the warm-up duration t_w with the *Time-based Ageing* (Sect. 5.2): In order to make the results comparable the *ageing time unit* was set to one minute ($t_{ur} = 1min$) and the simulation produced on average approximately one instance per minute (instance occurrence: $1/min$; deviation: 0.5). As a result similar t_{if} for the occurrence-based and time-based[6] ageing should yield results of a similar magnitude and are thus comparable, i.e. 10 traces $\approx$ 10 min. In Fig. 9 the development of warm-up duration t_w for the DCCM with *Time Ageing* for different $m \in \{1,5\}$ and $t_{if} \in \{0.005, 0.01, 0.02, 0.05, 0.1\}$ is depicted. As a reference the results of the *Occurrence Ageing* with $m = 1$ were also added to the graph. Similar to the occurrence-based ageing, the development of the warm-up duration t_w seems to be in no relation to the trace influence factor t_{if} for the time-based ageing. This indicates that the optimisation rule for the time-based ageing successfully improves and nullifies the default ageing method for the warm-up phase (see Eqs. 3 and 4 in Sect. 5.2). Additionally, it can be seen that the result of the two different ageing types with a similar configuration yield similar results (for $m = 1$).

In order to examine the effect of the optimisations for both of the ageing methods, the same experiments were repeated without the respective optimisations (only for $m = 1$). Figure 10 shows that especially for the occurrence ageing a sizeable improvement was achieved through the application of the proposed optimisations, e.g. for $t_{if} = 0.02$ the non-optimized occurrence ageing took on average 341 traces until the model was rediscovered but the optimised version was already successful after 8.44 traces (on average). The non-optimised

---

[6] Time-based ageing is technically based on an *ageing rate* a_r rather than *trace influence factor* t_{if}. However, a_r can be derived from t_{if}, i.e. $a_r = 1 - t_{if}$.

time-based ageing already yielded comparatively good results which could be further improved by the optimisation proposed (see Eq. 3 in Sect. 5.2), e.g. for $t_{if} = 0.02$ the non-optimized time ageing took on average 37.16 min ($\approx$ number of traces) until the model was rediscovered but the optimised version was already successful after an average of 8.42 min.

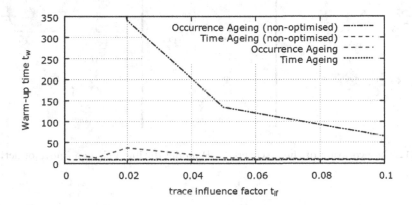

Fig. 10. Warm-up time without optimisation in relation to the trace influence factor with $m = 1$

6.3   Change Evaluation

In a second experiment we applied three changes of different extent to the business process (moving of an activity, adding an activity, and a complete process swap) during execution and are interested in the behaviour of the DCCM as well as in the change detection t_d and the change transition period t_{tr}.

Moving of Activity "A". The change applied is the move of activity A from the position before the inner Parallel construct to the position behind it. Figure 11 shows the evolution of the discovered BP models with occurrence ageing and *trace influence factor* $t_{if} = 0.01$ and *interpretation frequency* $m = 10$. The change was applied after 5753 traces. The footprint interpretation detects at the first chance to discover the change (trace 5760) a concept drift and finds via competition the best fitting construct: Parallel of a, c and b, d. The change detection t_d seemed to be unrelated to m and t_{if} for all experiment runs and was immediately recognized every time[7]. In Fig. 12 the development of t_{tr} for both ageing approaches different $m \in \{1, 10\}$ and $t_{if} \in \{0.005, 0.01, 0.02, 0.05, 0.1\}$ is shown. For this change a clear difference between the performance of the occurrence-based and the time-based ageing can be observed: The time-based

---

[7] Note, that other changes like deletion of an activity will take longer to recognise, since their existence still "lingers" in the footprints "memory" for some time.

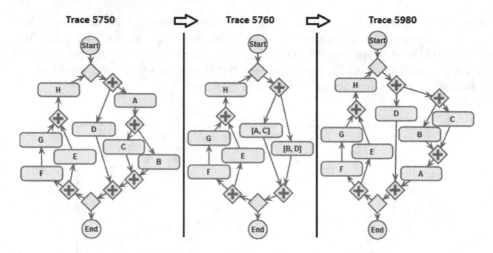

Fig. 11. The evolution of the discovered BP model during a change (move of activity "A")

ageing is significantly slower to finally detect the correct changed process, e.g. for $t_{if} = 0.02$ and $m = 1$ the time-based method detects the correct process BP_2 after on average 345 traces/minutes while the occurrence-based method recognises the new process correctly already after 105 traces (on average). Furthermore, it is observable that the change transition period t_{tr} was not particularly influenced by the interpretation frequency m but strongly influenced by t_{if}. If the value was very small ($t_{if} = 0.005$) a change took on average 450 traces (occurrence-based) or 1382 traces/minutes (time-based) in order to be reflected correctly in the discovered BP model. On the other hand if the trace influence factor is chosen very high, e.g. $t_{if} = 0.1$, the new process is correctly discovered after 21.1 (occurrence-based) or 67 (time-based) traces/minutes.

Adding of Activity "I". The change applied in this scenario is the addition of activity I at the end of the process. Figure 13 shows the evolution of the discovered BP models with occurrence ageing and *trace influence factor* $t_{if} = 0.01$ and *interpretation frequency* $m = 10$ while the change was applied after 5753 traces. It can be observed that the change detection t_d was again immediate, i.e. unrelated to m and t_{if}. However, it took on average 140 traces longer for the transition phase to be completed than for the previous scenario, i.e. on average 690 traces were necessary. The intermediate model which was valid from trace 5760 until 6450 (exclusively) recognises the relative position of activity I correctly (at the end of the process) but makes the activity optional. This is due to the fact that the "memory" of the dynamic overall footprint still contains behaviour from the original process in which the process ended without an activity I. Only after a certain amount of traces (dependent on the trace influence factor t_{if}) the memory of this behaviour became insignificant. Figure 14 shows an overview of the development of t_{tr} for all changes and both ageing approaches with $m = 1$ and $t_{if} \in \{0.005, 0.01, 0.02, 0.05, 0.1\}$. Both ageing approaches behave similarly for the

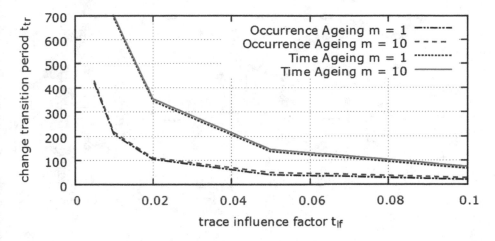

Fig. 12. The change transition period in relation to the trace influence factor (for recognising move of activity "A")

change of introducing the new activity I: even though the time-based ageing was usually slightly quicker, e.g. for $t_{if} = 0.01$: 687 (occurrence-based) vs. 684 (time-based) traces/minutes, the transition periods essentially only differed marginally (2 to 6 traces). When comparing the transition period t_{tr} for the occurrence-based method we can conclude that the movement of an activity was quicker detected correctly than the addition of a new activity ($t_{if} = 0.01$: 209 (move A) vs. 687 (add I)). However, for the time-based ageing only a relatively small difference was observed ($t_{if} = 0.01$: 691 (move A) vs. 684 (add I)).

Complex Change (Swap of Complete Process). The change applied in third scenario is a complete exchange of the original process during runtime, i.e. a revolutionary change. Figure 15 shows the evolution of the discovered BP models with occurrence ageing and *trace influence factor* $t_{if} = 0.01$ and *interpretation frequency* $m = 10$ with the change being applied after 5753 traces. The change detection t_d was again immediate (not displayed in Fig. 15), thus being unrelated to m and t_{if}. Many different process versions occurred during the transition phase of which only the version at trace 6310 is exemplary shown in the figure. Process BP_2 which is extremely different from BP_1 was finally correctly detected at trace 6680, i.e. on average 927 traces after the change was applied and 230 traces later than the introduction of a new activity (second scenario). When comparing the performance of occurrence-based and time-based ageing it can be observed that both develop similarly in relation to the trace influence factor t_{if} (see top two graphs in Fig. 14). This scenario can be considered the baseline scenario for how long a the transition phase may last at maximum for any t_{if}.

6.4   Occurrence-Based vs. Time-Based Ageing

A first observation is that the change detection t_d for all changes and ageing configuration was always quasi-immediate, i.e. whenever the first interpretation

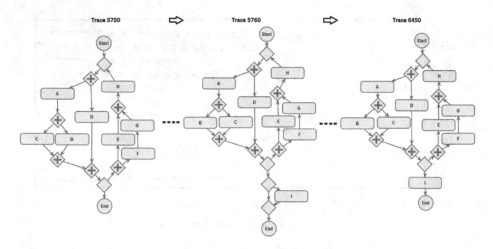

Fig. 13. The evolution of the discovered BP model during a change (addition of activity "I")

occurred after a change it was detected that the process has changed. This is mainly due to the configuration of the core CCM component and may change if different interpretation thresholds are selected. The ageing strategy, however, seems to not influence this directly (unless an extremely low trace influence t_{if} is selected).

A second observation is that there is a significant difference for the transition duration t_{tr} depending on the extent of the change, e.g. the movement of activity A took on average only $t_{tr} = 209$ traces to be correctly recognised whereas swapping process BP_1 with an entirely different process took on average $t_{tr} = 932$ to be recognised by the DCCM with occurrence-based ageing and $t_{if} = 0.01$ and $m = 1$.

Thirdly, time-based and occurrence-based ageing perform in most cases similarly well, with the exception of change scenario 1, in which activity A was moved to a different position within the process. Here, the occurrence-based ageing was able to detect the change significantly earlier than the time-based method, e.g. for $t_{if} = 0.1$ the new process is correctly discovered after 21.1 (occurrence-based) or 67 (time-based) traces/minutes. However, a fact not shown in the graphs is the number of exceptional results: While the time-based ageing did not suffer any exceptional experiment results it was observed that in approximately 4 % of the experiments for the occurrence-based method disproportionately high values for t_{tr} occurred. Due to the experiment setup, these results were ignored (see Sect. 6.1 - removal of the five highest and lowest values). In this context it was furthermore observed that a very high trace influence $t_{if} >> 0.1$ (not part of the above experiments) resulted in frequently changing/alternating discovered BP models even though the source process producing the events did not change. The reason for this behaviour is that not all variations of the process are included in the current dynamic footprint because they have already been

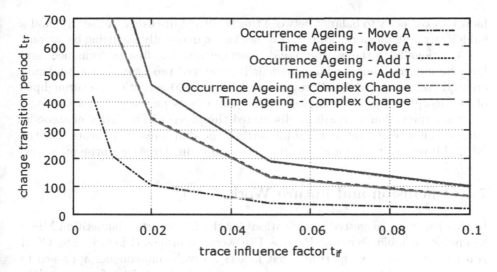

Fig. 14. The change transition period in relation to the trace influence factor for all three changes $(m = 1)$

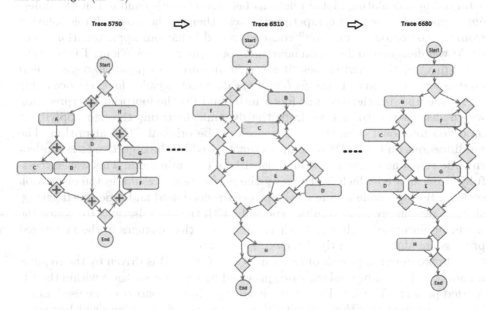

Fig. 15. The evolution of the discovered BP model during a change (complex change)

"forgotten" before they reappeared. It was observed that for the examined scenarios the occurrence-based ageing is slightly more susceptible to this issue than the time-based approach. Generally, the issue of "forgetting" too quickly is more likely to occur in large business processes containing rarely executed but still relevant behaviour and emphasises the importance of setting the trace influence factor t_{if} correctly to balance between timely correct discovery (higher t_{if}) and a sufficiently long memory (lower t_{if}) to not forget frequently occurring behaviour.

Additionally to the warm-up and change evaluation, first performance tests have been carried out for large artificially produced processes. For a randomly created and strongly nested process consisting of 100 activities the throughput of the footprint update was close to 100,000 events per second and the footprint interpretation successfully discovered the process in a matter of seconds. Although not tested yet in a real-life setting, the shown results indicate that the DCCM is very suitable for discovering and monitoring large enterprise processes.

7   Conclusion and Future Work

In this paper we suggested modifications for the Constructs Competition Miner to enable Scalable Dynamic Process Discovery as proposed in [16]. The CCM is a process discovery algorithm that follows a divide-and-conquer approach to directly mine a block-structured process model which consists of common BP-domain constructs and represents the main behaviour of the process. This is achieved by calculating global relations between activities and letting the different supported constructs compete with each other for the most suitable solution from top to bottom using "soft" constraints and behaviour approximations. The CCM was designed to deal with noise and not-supported behaviour. To apply the CCM in a real-time environment it was split up into two separate parts, executed on different occasions: (1) the *footprint update* which is called for every occurring event and updates the dynamic footprint(s) and (2) the footprint interpretation which derives the BP model from the dynamic footprint through applying a modified top-down competition approach of the original CCM algorithm. The modifications on the CCM were mostly motivated by the objective to keep algorithmic run-time of the individual algorithms to a minimum. This was successfully implemented which is shown by the performance results in the evaluation section. Both possible ageing methods, occurrence-based and time-based ageing, showed reasonably good results, especially with the optimisations to reduce the warm-up duration t_w. It was furthermore shown that changes in the monitored process are almost instantly detected, i.e. $t_d \approx 0$.

The presented approach of Dynamic CCM (DCCM) is driven by the requirements of real life industrial use cases provided by business partners within the EU funded project TIMBUS. During the evaluation in the context of the use-cases it became apparent that this concept still has a number of limitations which are considered to be future work: (1) Changes in the state of the business process are usually detected almost immediately but it may take a long time until the new state of the system is reflected appropriately in the extracted business process model. This behaviour originates from the fact that the footprint and the interpreted business process are in a sort of intermediate state for a while until the influence of the old version of the business process has disappeared. Furthermore, the trace influence factor t_{if} is a pre-specified value but in reality it is dependent on how many traces we need to regard to represent all the "behaviour" of the model[8]. This in turn is strongly dependent on the amount of activities in the model, since more activities usually mean more control-flow behaviour. A possible future modification could be to have the influence factor dynamically adapt, i.e. similar to the self-adapting ageing proposed in [5]. (2) If no sub-footprint is available for a set of activities, the footprint interpreter does not further analyse this set. Through approximations or the use of the direct neighbours relation at least a "close enough" control-flow for the subset could be retrieved. (3) The discovery of the state of a business process should also comprise information of other perspectives than the control-flow, e.g. resource and performance.

References

1. von Ammon, R., Ertlmaier, T., Etzion, O., Kofman, A., Paulus, T.: Integrating complex events for collaborating and dynamically changing business processes. In: Dan, A., Gittler, F., Toumani, F. (eds.) ICSOC/ServiceWave 2009. LNCS, vol. 6275, pp. 370–384. Springer, Heidelberg (2010)
2. Bose, R.P.J.C., van der Aalst, W.M.P., Žliobaitė, I., Pechenizkiy, M.: Handling concept drift in process mining. In: Mouratidis, H., Rolland, C. (eds.) CAiSE 2011. LNCS, vol. 6741, pp. 391–405. Springer, Heidelberg (2011)
3. Cattafi, M., Lamma, E., Riguzzi, F., Storari, S.: Incremental declarative process mining. In: Szczerbicki, E., Nguyen, N.T. (eds.) Smart Information and Knowledge Management. SCI, vol. 260, pp. 103–127. Springer, Heidelberg (2010)
4. Buijs, J., Van Dongen, B., Van Der Aalst, W.: A genetic algorithm for discovering process trees. In: Evolutionary Computation (CEC), pp. 1–8. IEEE (2012)
5. Burattin, A., Sperduti, A., Van Der Aalst, W.: Heuristics miners for streaming event data. CoRR abs/1212.6383 (2012)
6. Kindler, E., Rubin, V., Schäfer, W.: Incremental workflow mining based on document versioning information. In: Li, M., Boehm, B., Osterweil, L.J. (eds.) SPW 2005. LNCS, vol. 3840, pp. 287–301. Springer, Heidelberg (2006)
7. Kindler, E., Rubin, V., Schäfer, W.: Activity mining for discovering software process models. In: Software Engineering 2006, Fachtagung des GI-Fachbereichs Softwaretechnik, LNI, pp. 175–180. GI (2006)
8. Ko, R.K.L.: A computer scientist's introductory guide to business process management (BPM). ACM Crossroads J. 15(4), 11–18 (2009)
9. Leemans, S.J.J., Fahland, D., van der Aalst, W.M.P.: Discovering block-structured process models from event logs - a constructive approach. In: Colom, J.-M., Desel, J. (eds.) PETRI NETS 2013. LNCS, vol. 7927, pp. 311–329. Springer, Heidelberg (2013)
10. Leemans, S.J.J., Fahland, D., van der Aalst, W.M.P.: Discovering block-structured process models from event logs containing infrequent behaviour. In: Lohmann, N., Song, M., Wohed, P. (eds.) BPM 2013 Workshops. LNBIP, vol. 171, pp. 66–78. Springer, Heidelberg (2014)
11. Luckham, D.: The Power of Events: An Introduction to Complex Event Processingin Distributed Enterprise Systems. Addison-Wesley Professional, Reading (2002)

---

[8] if t_{if} is set too high normal behaviour unintentionally becomes exceptional behaviour.

12. Maggi, F.M., Burattin, A., Cimitile, M., Sperduti, A.: Online process discovery to detect concept drifts in LTL-based declarative process models. In: Meersman, R., Panetto, H., Dillon, T., Eder, J., Bellahsene, Z., Ritter, N., Leenheer, P., Dou, D. (eds.) ODBASE 2013. LNCS, vol. 8185, pp. 94–111. Springer, Heidelberg (2013)

13. Manku, G.S., Motwani, R.: Approximate frequency counts over data streams. In: VLDB 2002, Proceedings of 28th International Conference on Very Large Data Bases, pp. 346–357. Morgan Kaufmann (2002)

14. OMG Inc.: Business Process Model and Notation (BPMN) Specification 2.0. formal/2011-01-03 (2011). http://www.omg.org/spec/BPMN/2.0/PDF

15. Redlich, D., Molka, T., Gilani, W., Blair, G., Rashid, A.: Constructs competition miner: process control-flow discovery of BP-domain constructs. In: Sadiq, S., Soffer, P., Völzer, H. (eds.) BPM 2014. LNCS, vol. 8659, pp. 134–150. Springer, Heidelberg (2014)

16. Redlich, D., Gilani, W., Molka, T., Drobek, M., Rashid, A., Blair, G.: Introducing a framework for scalable dynamic process discovery. In: Aveiro, D., Tribolet, J., Gouveia, D. (eds.) EEWC 2014. LNBIP, vol. 174, pp. 151–166. Springer, Heidelberg (2014)

17. Redlich, D., Blair, G., Rashid, A., Molka, T., Gilani, W.: Research challenges for business process models at run-time. In: Bencomo, N., France, R., Cheng, B.H.C., Aßmann, U. (eds.) Models@run.time. LNCS, vol. 8378, pp. 208–236. Springer, Heidelberg (2014)

18. Solé, M., Carmona, J.: Incremental process mining. In: Proceedings of the Workshops of the 31st International Conference on Application and Theory of Petri Nets and Other Models of Concurrency (PETRI NETS 2010) and of the 10th International Conference on Application of Concurrency to System Design (ACSD 2010). CEURWorkshop Proceedings, pp. 175190. CEUR-WS.org (2010)

19. Van Der Aalst, W., Ter Hofstede, A.: YAWL: Yet Another Workflow Language (2003)

20. Van Der Aalst, W., Weijters, A., Maruster, L.: Workflow mining: discovering process models from event logs. IEEE Trans. Knowl. Data Eng. **16**(9), 1128–1142 (2004)

21. Van Der Aalst, W., Van Dongen, B.: ProM: the process mining toolkit. Ind. Eng. **489**, 1–4 (2009)

22. van der Aalst, W., et al.: Process mining manifesto. In: Daniel, F., Barkaoui, K., Dustdar, S. (eds.) BPM Workshops 2011, Part I. LNBIP, vol. 99, pp. 169–194. Springer, Heidelberg (2012)

23. Van Der Aalst, W.: Process Mining - Discovery, Conformance and Enhancementof Business Processes. Springer, Heidelberg (2011)

24. Van Der Aalst, W., Adriansyah, A., Van Dongen, B.: Replaying history on process models for conformance checking and performance analysis. WIREs Data Min. Knowl. Discov. **2**(2), 182–192 (2012)

25. Weijters, A., Van Der Aalst, W., Alves de Medeiros, A.: Process mining with the heuristics miner-algorithm. BETA Working Paper Series, WP 166, Eindhoven University of Technology (2006)

Trustworthy Cloud Certification: A Model-Based Approach

Marco Anisetti[1], Claudio A. Ardagna[1], Ernesto Damiani[1], and Nabil El Ioini[2(✉)]

[1] Università degli Studi di Milano, DI, Crema, Italy
{marco.anisetti,claudio.ardagna,ernesto.damiani}@unimi.it
[2] Free University of Bozen, Bolzano, Italy
nabil.elioini@unibz.it

Abstract. Cloud computing is introducing an architectural paradigm shift that involves a large part of the IT industry. The flexibility in allocating and releasing resources at runtime creates new business opportunities for service providers and their customers. However, despite its advantages, cloud computing is still not showing its full potential. Lack of mechanisms to formally assess the behavior of the cloud and its services/processes, in fact, negatively affects the trust relation between providers and potential customers, limiting customer movement to the cloud. Recently, cloud certification has been proposed as a means to support trustworthy services by providing formal evidence of service behavior to customers. One of the main limitations of existing approaches is the uncertainty introduced by the cloud on the validity and correctness of existing certificates. In this paper, we present a trustworthy cloud certification approach based on model verification. Our approach checks certificate validity at runtime, by continuously verifying the correctness of the service model at the basis of certification activities against real and synthetic service execution traces.

Keywords: Certification · Cloud · FSM · Model verification

1 Introduction

Cloud computing paradigm is radically changing the IT infrastructure, as well as traditional software provisioning and procurement. Cloud-based services are becoming the primary choice for many industries due to the advantages they offer in terms of efficiency, functionality, and ease of use. Several factors characterize the choice between functionally-equivalent services at infrastructure, platform, and application layers, among which Quality of Service (QoS) stands out [17]. However, the dynamic and opaque nature of the cloud makes it hard to preserve steady and transparent QoS, which often affects the trust relationship between providers and their customers, and limits customer movement to the cloud.

In response to the need of a trustworthy and transparent cloud environment, several assurance techniques have been defined [6,12,16,22,23]. Among them,

© IFIP International Federation for Information Processing 2015
P. Ceravolo et al. (Eds.): SIMPDA 2014, LNBIP 237, pp. 107–122, 2015.
DOI: 10.1007/978-3-319-27243-6_5

certification approaches are on the rise [2,10,26]. Cloud certification in fact is beneficial for both customers having trusted evidence on the correct behavior of the cloud (and corresponding cloud/service providers) in the treatment and management of their data and applications, and providers having trusted evidence on the truthfulness of their claims.

Software certification has a long history and has been used in several domains to increase trust between software system providers and customers. A certificate allows providers to gain recognition for their efforts to increase the quality of their systems, and supports trust relationships grounded on empirical evidence [26]. However, cloud computing introduces the need of re-thinking existing certification techniques in light of new challenges, such as services and processes owned by unknown parties, data not fully under control of their owners, and an environment subject to rapid and sudden changes. Certification processes need then to be tailored to accommodate these new challenges.

An increasing trend in software system certification is to test, monitor, and/or check a system behavior according to its model [4,7,24]. If this paradigm fits well the certification of a static system, it opens the door to potential inconsistencies in cloud environments, where cloud services and corresponding processes are subject to changes when deployed in the production environment, and could therefore differ from their counterparts verified in a lab environment. In this scenario, online certification is a fundamental requirement and evidence collection becomes a continuous and runtime process. In general we need to answer the following question: *"does my service behave as expected at runtime when deployed in the production environment?"*. The correct answer to this question passes from the continuous verification of the model used to provide a solid evidence on service behavior. In this paper, we present an approach to continuous model verification at the basis of a sound cloud service certification. Our approach builds on testing and monitoring of execution traces to discover differences between the model originally used to verify and certify a service in a lab environment, and the one inferred in the production environment.

The remainder of this paper is structured as follows. Section 2 presents our certification process and reference scenario. Section 3 presents two different approaches to model generation. Section 4 describes our approach to model verification. Section 5 presents a certification process adaptation based on model verification in Sect. 4. Section 6 illustrates an experimental evaluation of our approach. Section 7 discusses related work and Sect. 8 draws our concluding remarks.

2   Certification Process and Reference Scenario

We describe the certification process at the basis of the approach in this paper and our reference scenario.

2.1   Certification Process

A certification process for the cloud involves three main parties [2], namely a *cloud/service provider* (service provider in the following), a *certification authority,*

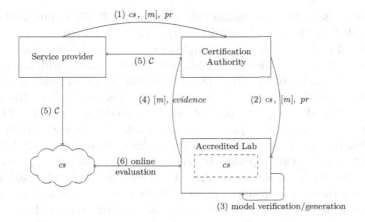

Fig. 1. Certification process

and an *accredited Lab*. It is aimed at collecting the evidence proving a property pr for a cloud service cs (Target of Certification – ToC). The process, being defined for the cloud, must combine both *offline* and *online evaluation*. Offline evaluation usually considers a copy of the ToC deployed in a lab environment, which can be tested according to the selected property. While offline evaluation verifies services in a controlled setting and provides many advantages such as improved performance and reduced costs, it does not represent alone a suitable approach for the cloud. For this reason, online evaluation has attracted increasing attention, since it provides rich information coming from real services and processes running with production configurations. It also permits to study the behavior of services at runtime, and verify that service certificates are valid over time and across system/environment changes. Online evaluation involves the real ToC, which can be evaluated by monitoring real traces of execution or by testing specific aspects of its behavior.

Figure 1 shows our certification process, taken as reference in this paper. A service provider initiates the certification process by requesting a certificate for one of its cloud services cs (ToC) for a given property pr (step 1). We note that the request *may* contain the model m of the service to be certified ([m] in Fig. 1). The certification authority receiving the request starts the process by involving its accredited lab (step 2). The latter is delegated by the certification authority to carry out the service evaluation. The accredited lab checks the correctness of model m (step 3), if available, or generates it according to information (e.g., interface description, implementation documentation, source code) available on the service. Model m is then used to collect the evidence, which is delivered to the certification authority (step 4). The certification authority evaluates whether the evidence is sufficient or not to prove the requested property for the service and issue a certificate $\mathcal{C}$ to it (step 5). Upon certificate issuing (step 5), the accredited lab starts a continuous and online evaluation process (step 6), involving model verification and refinement (step 3). We note that, although some solutions for online evaluation exist [5,27], online evaluation is often preceded by offline evaluation to target those scenarios which are difficult to evaluate on real systems (e.g., Denial of Service, security attacks).

In summary, we consider a certification process whose evidence collection is driven by a model of the ToC [2]. The model is an automaton representing the ToC behavior. It is used to exercise the ToC and collect the evidence needed to verify the property of interest. The model is verified both offline, when provided by the service provider, and online, where its validity is continuously evaluated together with the validity of the corresponding certificate. Our certification process can support different types of evidence (e.g., test-based, monitoring-based) to assess the quality of service. On one hand, test-based certification is grounded on results retrieved by test case executions on the ToC, while monitoring-based certification builds on evidence retrieved by observing real executions of the ToC. The choice of defining a cloud certification scheme based on system modeling is driven by the fact that model-based approaches are common in the context of software evaluation [2,7,19,20,24]. In particular, model-based approaches have been used to analyze software behavior, to prove non-functional properties of software, to infer system executions, and to generate test cases at the basis of software evaluation.

2.2   Reference Scenario

Our reference scenario is a travel planner *TPlan* delivered at cloud application layer (SaaS),[1] implementing functionality for flight and hotel reservation. It is implemented as a composite service and includes three main parties: *(i)* the client requesting a travel plan; *(ii)* the travel agency (i.e., *TPlan*) implementing service *TPlan* and acting as the service orchestrator; and *(iii)* the component services which are invoked by the travel agency within *TPlan* process. The travel agency implements *TPlan* as a business process using a service *BookFlight* for flight reservation, a service *BookHotel* for hotel reservation, and a service *Bank-Payment* for payment management. Table 1 summarizes the details about the operations of partner services, including *TPlan*.

Upon logging into the system by means of a public interface provided by *TPlan* (operation login), customers submit their preferences, which are distributed to the partner operations findOffers of services *BookFlight* and *BookHotel*. Once the customer has selected the preferred flight and hotel (calling operations bookOffer of services *BookFlight* and *BookHotel*, respectively), service *BankPayment* is invoked to conclude the reservation (operation makePayment of service *BankPayment*). We note that *BankPayment* is invoked only in case both *BookFlight* and *BookHotel* are correctly executed. The customer can also cancel a previous transaction and logout from the system (operations cancelTPlan and logout of *TPlan*).

---

[1] We note that, though for simplicity a SaaS scenario is considered in the paper, the proposed approach applies to services insisting also on platform (PaaS) and infrastructure (IaaS) layers.

Table 1. Operations of service travel planner

Service	Operation	Description
TPlan	< tokenID > login(*username, password*)	Provides password-based authentication, and returns an authentication token
TPlan	< TPlanID > saveTPlan(*hotelBookings, flightBookings*)	Saves hotel and flight reservations
TPlan	< confirmation > cancelTPlan(*TPlanID*)	Cancels a plan
TPlan	< confirmation > logout (*tokenID*)	Disconnects a user and destroys the authentication token
BookHotel	<confirmation> login (*tokenID*)	Provides a token-based authentication
BookHotel	< hotelsList > findOffers(*check-in, check-out*)	Searches for hotel offers and returns a list of offers
BookHotel	< confirmation > bookOffer (*offerID*)	Books a specific offer from the offer list
BookHotel	< confirmation > cancel(*bookingID*)	Cancels an existing reservation
BookHotel	< confirmation > logout(*tokenID*)	Disconnects a user from service BookHotel
BookFlight	< confirmation > login(*tokenID*)	Provides a token-based authentication
BookFlight	< flightsList > findOffers(*departure-day, return-day*)	Searches for flight offers and returns a list of offers
BookFlight	< confirmation >bookOffer(*offerID*)	Books a specific offer from the offer list
BookFlight	< confirmation > cancel(*bookingID*)	Cancels an existing reservation
BookFlight	< confirmation >logout(*tokenID*)	Disconnects a user from service BookFlight
BankPayment	< transactionID >makePayment(*tokenID, TPlansID*)	Executes a payment
BankPayment	< confirmation > cancelPayment(*transactionID*)	Cancels a transaction

3   System Modeling

The trustworthiness of a model-based cloud certification process is strictly intertwined with the trustworthiness of the considered model, or in other words depends on how much the model correctly represents the ToC. The latter depends on *(i)* the amount of information available on the system to be modeled and *(ii)* how the model is generated. We consider two types of models depending on the amount of available information: *(i) workflow-based models* that consider information on service/operation conversations, *(ii) implementation-based models* that extend workflow-level models with details on operation and service implementation. In the following, we give a quick overview on workflow-based and implementation-based models.

3.1   Workflow-Based Model

A workflow-based model represents the ToC in terms of operations/services. At this level we differentiate between two types of workflow: *(i) single-service workflow*, which models the sequence of operation invocations within a single service and *(ii) composite-service workflow*, which models a sequence of operation invocations within a composite service. We note that, in the latter case, operations belong to different composed services. We also note that single-service work-flow can be considered as a degeneration of the general case of composite-service workflow, modeling the internal flow of a single service. Some approaches already used workflow-based models for service verification (e.g., [15,20]). Merten et al. [20] presented a black-box testing approach for service model generation, which entirely relies on service interface and addresses requirements of single-service workflow. Their approach focuses on the definition of a data-sensitive behavioral model in three steps as follows. First, it analyzes the service interface and generates a dependency automaton based on input/output dependencies between operations. Second, a saturation rule is applied, adding possible service invocations from the client to the model (e.g., directly calling an operation without following the subsequent calls). Third, an additional step verifies whether the generated dependencies are semantically meaningful or not. Fu et al. [15] addressed the issue of correctness verification of composite services using a model based approach. To this aim, they developed a tool to check that web services satisfy specific Linear temporal logic properties. Their approach relies on BPEL specifications, which are translated in guarded automata. Automata are then translated into Promela language and checked using the SPIN model checker.

3.2 Implementation-Based Model

An implementation-based model extends the workflow-based model with implementation details coming from the service providers. These details can be used in different ways depending on the type of information they carry on. If they include implementation documentation, they can be used to manually build an automaton representing the behavior of the single operations. Otherwise, if the providers provide traces of the internal operation execution, methods such as the one in [19] can be used to automatically generate the internal behavioral model of the service. Each of the states in the workflow-based model can be further extended in the implementation-based model. At this level we can combine techniques for extracting the service model based on data value constraints [14] and techniques that generate a finite state machine based on service component interactions [8].

Example 1. Figure 2 presents a *workflow-based model* (*composite-service*), a *workflow-based model* (*single service*), and an *implementation-based model*. Workflow-based model (composite-service) is driven by the flow of calls between services in the composition. For instance, *BookFlight* and *BookHotel* are two partner services that are invoked by *TPlan*. *Workflow-based model* (*single service*) is driven by the flow of operation calls within a service, which are annotated over the arcs. For instance, service *BookHotel* exposes different operations to book a hotel room (see Table 1), which are used to generate its FSM model [20]. The model includes the input/output dependencies between operations. Also, for simplicity, a transition is triggered iff the execution of the annotated operation call is successful. Node *Env* in Fig. 2 represents requests sent by the client to different operations. *Implementation-based model* is driven by the flow of code

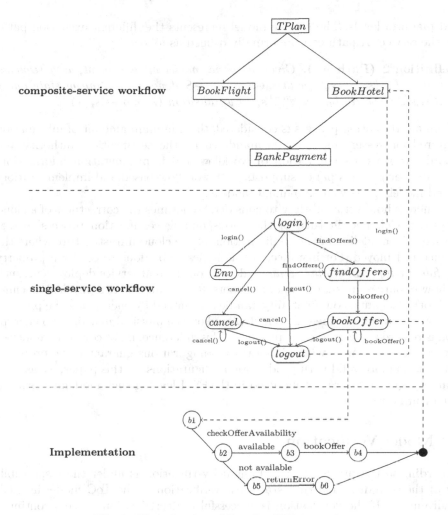

Fig. 2. Three-level modeling for *TPlan* service

instructions within a service operation and provides a fine-grained representation of the considered service.

All generated models can be represented as a Finite State Machine (FSM) defined as follows.

Definition 1 (Model m). *Let us consider a service model m generated by the accredited lab. Model m can be defined as $m=(S, \Sigma, \delta, s_0, F)$, where S is a finite set of states, Σ is a finite set of input, $\delta : S \times \Sigma \mapsto S$ is the transition function, $s_0 \in S$ is the initial state, and $F \subseteq S$ is the set of final states.*

We note that each transition annotation in Σ corresponds to a valid operation/code instruction call at any levels of the cloud stack, including infrastructure and platform levels. The service model represents the different execution paths for the service. A path can be formally defined as follows.

Definition 2 (Path pt_i). *Given a service model m, a path pt_i is a sequence of states $pt_i = \langle s_0, \ldots, s_n \rangle$, with $s_0 \in S$ and $s_n \in S$ denoting the initial state and a final state, respectively, s.t. $\forall_{i=0}^{n-1} s_i$, $\exists$ a transition $(s_i \times \sigma \mapsto s_{i+1}) \in \delta$.*

When a certification process is considered, the minimum amount of information required for system modeling is mandated by the certification authority and usually involves a combination of workflow and implementation information. The approach in this paper supports both workflow-based and implementation-based modeling, as well as manual modeling.

There is however a subtlety to consider. Sometimes the correctness of a cloud service model, and in turn of the corresponding certification process, passes through the modeling of the configurations of the cloud infrastructure where the service is deployed. For instance, the activities to be done to certify a property *confidentiality against other tenants* depend on the real service deployment, such as how resources are shared among tenants. If a tenant shares a physical machine with other tenants, confidentiality can be guaranteed by encrypting the physical storage; if a tenant is the only one deployed on the physical machine, no encryption is required to preserve the property. This difference, if not correctly modeled, can result in scenarios where laboratory configurations guarantee the property, which is no more valid with production configurations. In this paper, we assume that configurations are not modeled in the FSM leaving such modeling issue for our future work.

4   Model Verification

According to the process in Fig. 1, model verification is under the responsibility of the accredited lab that starts the verification of the ToC model in a lab environment. If the verification is successful the certification process continues and eventually leads to certificate issuing based on the collected evidence. We note that, in those cases where the model is generated by the accredited lab itself, model verification is successful by definition. Upon certificate issuing, the certificate is linked to the ToC deployed in the production environment. However, in addition to common errors that could inadvertently be added during model definition, the ToC can be affected by events changing its behavior once deployed in the cloud production environment, and therefore impairing the correctness of the original modeling effort usually done in a lab environment. It is therefore fundamental to provide an approach to online model verification, which allows to continuously evaluate the correctness and trustworthiness of certification processes and the validity of the corresponding issued certificates. The accredited lab is responsible for online model verification, and checks the consistency between the observed ToC behavior and the original model.

In the following of this section, we present our model verification approach. The approach checks model correctness by collecting execution traces from real

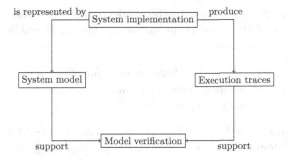

Fig. 3. Model verification process

ToC executions and projecting them on the model itself. Figure 3 shows the conceptual design of our approach.

4.1 Execution Trace Collection

Execution trace collection represents the first step of our verification process. Traces are collected either by monitoring real executions of the ToC involving real customers or by observing the results of ad hoc test-based synthetic executions (synthetic traces). Execution traces can be formally defined as follows.

Definition 3 (Trace T_i). *An execution trace T_i is a sequence $\langle t_1,\ldots,t_n \rangle$ of actions, where t_j can be either an operation execution op_j or a code instruction execution ci_j.*

We note that a trace T_i composed of a sequence of operation executions op_j refers to a workflow-based model, while a trace T_i also including code instruction executions ci_j refers to an implementation-based model. We also note that execution traces can be collected at multiple provider sites, depending on the considered workflow.

Example 2. According to our reference scenario in Sect. 2.2, two types of traces can be collected depending on the considered level (see Fig. 2). At workflow level, a trace represents a sequence of operation invocations, which may belong to a single (e.g., $T_i = \langle BookHotel.\texttt{login}(), BookHotel.\texttt{findOffers}(), BookHotel.\texttt{bookOffers}()\rangle$) or multiple (e.g., $T_i = \langle BookHotel.\texttt{login}(), BookHotel.\texttt{findOffers}(), BookHotel.\texttt{bookOffers}(), BankPayment.\texttt{makePayment}()\rangle$) services. At implementation level, a trace is represented by a sequence of code instruction executions (e.g. $T_i = \langle \texttt{checkOfferAvailability}, \texttt{notavailable}, \texttt{returnError}\rangle$).

4.2 Service Model Verification

Model verification is a function that takes as input the service model $m = (S, \Sigma, \delta, s_0, F)$ and collected execution traces T_i, and produces as output either *success* (1), if the traces conform to the service model, or *failure* (0), otherwise, with a description of the type of inconsistency found. Formally, we can define model verification as follows.

Definition 4 (*MV*). *Model Verification is a function* $MV : \mathcal{M} \times \mathcal{T} \rightarrow \{0,1\} \times \mathcal{R}$ *that takes as input the initial service model* $m = (S, \Sigma, \delta, s_0, F) \in \mathcal{M}$ *and execution traces* $T_i \in \mathcal{T}$, *and produces as output either:*

- $[1, \emptyset]$ *iff i)* $\forall T_i \in \mathcal{T}$, $\exists$ *a finite path* $pt_j = \langle s_0, \ldots, s_n \rangle$ *in* $m \in \mathcal{M}$ *s.t.* T_i *is consistent with* pt_j *(denoted* $T_i \equiv pt_j$) *and ii)* $\forall pt_j = \langle s_0, \ldots, s_n \rangle$ *in* $m \in \mathcal{M}$, $\exists T_i \in \mathcal{T}$ *s.t.* $T_i \equiv pt_j$, *or*
- $[0, r]$, *otherwise, where* $r \in \mathcal{R}$ *describes the reason why a failure is returned.*

We note that model verification is based on a *consistency function* $\equiv$ between collected traces and service model paths, as defined in the following.

Definition 5 (Consistency Function $\equiv$). *Given a trace* $T_i = \langle t_1, \ldots, t_n \rangle \in \mathcal{T}$ *and* $pt_j = \langle s_0, \ldots, s_n \rangle$ *in* $m \in \mathcal{M}$, $T_i \equiv pt_j$ *iff* $\forall t_k \in T_i$, $\exists (s_{k-1} \times \sigma \mapsto s_k) \in \delta$ *s.t.* t_k *and* σ *refer to the same operation/code instruction.*

Definitions 4 and 5 establish the basis for verifying the consistency between observed traces and paths. A failure in the model verification means that there is an inconsistency between the service model and the execution traces, which can affect an existing certification process and invalidate an issued certificate (see Sect. 5). Several types of inconsistencies can take place, which can be reduced to three main classes as follows.

- *Partial path discovery:* it considers a scenario in which a trace is consistent with a subset of a path in the model. In other words, given a trace $T_i = \langle t_1, \ldots, t_n \rangle \in \mathcal{T}$ and a path $pt_j = \langle s_0, \ldots, s_l \rangle$ in $m \in \mathcal{M}$, $\exists$ a subset $\overline{pt_j}$ of pt_j s.t. $T_i \equiv \overline{pt_j}$. This means that while mapping a trace to paths in the model, only an incomplete path is found. We note that, for traces that stop before a final state in m, an inconsistency is raised after a pre-defined timeout expires. The timeout models the expected elapsed time between two operations/instructions execution.
- *New path discovery:* it considers a scenario in which a trace is not consistent with any path in the model. In other words, given a trace T_i and a model m, $\forall pt_j \in m$, $T_i \not\equiv pt_j$. This means that a new path is found, that is, at least a new transition and/or a new state is found in the traces.
- *Broken existing path:* it considers a scenario in which real traces do not cover a path in the model, and the synthetic traces return an error for the same path. In other words, given a path pt_j, $\nexists T_i$ s.t. $T_i \equiv pt_j$. This means that the model includes a path that is not available/implemented in the real ToC.

We note that the above classes of inconsistency are due to either a bad modeling of the service or a change in the production service/environment. Broken existing path inconsistencies can also be due to unexpected failures within the deployed ToC. We also note that additional inconsistencies can be modeled by

Table 2. Execution traces of operation *BookHotel*.

# Trace	Trace values
1	login(),findOffers(),bookOffer(),logout()
2	bookOffer(),cancel()
3	findOffers(),login(),bookOffer(),logout()
4	Env,cancel()
5	Env,login(),logout()
6	Env,cancel(),login(),cancel(),logout()
7	Env,bookOffer(),login(),bookOffer(),logout()
8	bookOffer(),login(),bookOffer(),cancel()
9	findOffers(),bookOffer(),login(),bookOffer(),logout()
10	findOffers(),login(),logout()

a combination of the above three. As an example, let us consider a single operation annotating a transition in model m (e.g., login() of service *BookHotel*), which is updated to a new version with a new slightly different interface. In this case, two inconsistencies are raised. First, a new path is discovered such that it contains the new interface for operation login(); then, a broken existing path is discovered having the original operation login().

Example 3 Let us consider the execution traces in Table 2. By projecting the list of traces over the model in Fig. 2 (*single service workflow*), we find some inconsistencies. For instance, trace 2 shows a partial path inconsistency. The sequence of calls (*BookHotel*.bookOffer(), *BookHotel*.cancel()) maps to a sub-path of the model (*login, bookOffer, cancel*). Trace 10 shows a new path discovery inconsistency. The sequence of calls (*BookHotel*.findOffers(), *BookHotel*.login(), *BookHotel*.logout()) is supported by the service, while the model does not have this path (i.e., the model is missing a transition from node *findOffers* to node *login*). Finally, let us consider a scenario in which a failure in the authentication mechanism makes function login() unreachable. In this case, a broken existing path inconsistency is raised for each path involving function login().

5 Certification Process Adaptation

Inconsistencies raised by our model verification in Sect. 4 trigger a certification process adaptation, which could result in a certificate refinement. Certification process adaptation is executed during online evaluation by the accredited lab. It is aimed at incrementally adapting the certification process according to the severity of the model inconsistency, reducing as much as possible the need of costly re-certification processes [3].

The accredited lab, during online evaluation, collects the results of our model verification including possible inconsistencies. Then, it adapts the entire evaluation process following four different strategies.

– *No adaptation:* model verification raises negligible inconsistencies (e.g., a *partial path discovery* that does not affect the certified property). Accredited lab confirms the validity of the certificate.
– *Re-execution:* model verification raises one or more *broken existing path* inconsistencies. Accredited lab triggers additional testing activities by re-executing test cases on broken paths, to confirm the validity of the certificate.
– *Partial re-certification:* model verification raises critical inconsistencies (e.g., a *new path discovery* or a *partial path discovery* that affects the certified property). Accredited lab executes new evaluation activities, such as exercising the new portion of the system for certificate renewing.
– *Re-certification:* *re-execution* or *partial re-certification* fail and the accredited lab invalidates the certificate. A new certification process re-starts from step 2 in Fig. 1 by adapting original model m according to the model verification outputs (i.e., r in Definition 4).

The role of model verification is therefore twofold. On one side, it triggers *re-execution* of testing activities; on the other side, it adapts the model for partial or complete re-certification. In any case, model verification supports a trustworthy cloud certification process, where accidental and/or malicious modifications to certified services are identified well in advance, reducing the window of time in which an invalid certificate is perceived as valid.

6   Experimental Evaluation

We implemented a Java-based prototype of our model verification approach. The prototype is composed of two main modules, namely, *consistency checker* and *model adapter*. *Consistency checker* receives as input a service model and a set of real and synthetic traces,[2] and returns as output inconsistent traces with the corresponding type of inconsistency. *Model adapter* receives as input the results of the consistency checker and, according to them, generates as output a refined model.

To assess the effectiveness of our prototype, we generated an experimental dataset as follows. We first manually defined the correct implementation-based model m_{cs} of a generic service cs composed of 20 different paths; we then randomly generated 1000 inconsistent models by adding random inconsistencies (Sect. 4.2) to m_{cs}. Inconsistent models are such that 10 %, 20 %, 30 %, 40 %, or 50 % of the paths are different (e.g., missing, new) from the paths in m_{cs}. To simulate realistic customer behaviors, we extended m_{cs} by adding a probability P_i to each path $pt_i \in m_{cs}$, such that $\sum_i P_i = 1$. Probabilities are taken randomly from a normal probability distribution such that there exist few paths whose probability of being invoked tends to 0. Real and synthetic traces are then produced using m_{cs} extended with probabilities.

---

[2] We remark that synthetic traces are generated by ad hoc testing.

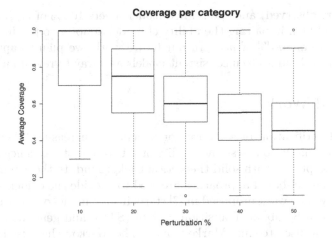

Fig. 4. Refined model coverage of m_{cs}.

The experimental evaluation proceeded as follows. First, using the *consistency checker*, we verified inconsistent models in the dataset and retrieved inconsistent traces together with the type of inconsistency. Then, using the *model adapter*, we built a refined model of each inconsistent model, and evaluated how much these models approximate the correct model m_{cs}. Our results show that the refined models covered 64 % of paths in m_{cs} on average, with an increase of 28 % on the average coverage of the inconsistent models. Also, 17 % of the inconsistent models were able to cover the entire model m_{cs} (i.e., 100 % coverage), while 0 % of the inconsistent models in the dataset covered the entire model by construction. Figure 4 shows a Box and Whisker chart presenting more detailed results on the basis of the rates of differences (i.e., 10 %, 20 %, 30 %, 40 %, 50 %) introduced in the inconsistent models. The Box and Whisker chart in Fig. 4 splits the dataset into quartiles. Within the box, containing two quartiles, the horizontal line represents the median of the dataset. Two vertical dashed lines, called whiskers, extend from the bottom and top of the box. The bottom whisker goes from box to the smallest non-outlier in the data set, and the top whisker goes from box to the largest non-outlier. The outliers are plotted separately as points on the chart. Figure 4 shows that while we increased the perturbation level, we decreased the ability to achieve a full coverage of the initial model. In fact, the median value decreases in such a way that the complete coverage is achieved in the last category (50 %) as an outlier. Nevertheless, in all the perturbation levels, 50 % of the models recover at least 40 % of the inconsistencies. Additionally, in the first case (10 % perturbation), more than 50 % of the models were completely recovered; therefore, both the median and the maximum coverage are equal to 1.

In summary, model adapter does not achieve full coverage of m_{cs} for all refined models. This is mainly due to the fact that paths at low probability are invoked with low probability. If these paths are already specified in an inconsistent model, then a synthetic trace can be generated to evaluate them (if no real traces are observed) and the refined model covers 100 % of m_{cs}. Otherwise, they remain hidden impairing the ability of model adapter to produce a refined model that covers 100 % of m_{cs}. In our future work, we plan to apply fuzzing and mutation techniques to inconsistent models as a way to reveal hidden paths.

7   Related Work

Model-based verification of software components is increasingly becoming the first choice technique to assess the quality of software systems, since it provides a systematic approach with solid theoretical background. In the context of cloud certification, model-based approaches are used to provide the evidence to certify service quality. The work proposed in [7] starts from a model of the service under certification as a Symbolic Transition System (STS), and generates a certification model as a discrete-time Markov chain. The Markov chain is then used to prove dependability properties of the service. In [4], the authors uses STSs to model services at different levels of granularity. The model is then used to automatically generate the test cases at the basis of the service certification process. Spanoudakis et al. [25] present the EU FP7 Project CUMULUS [9], which proposes a security certification scheme for the cloud based on the integration of different techniques for evidence collection and validation. Additionally, they support also an incremental approach to certify continuously evolving services. A different strategy proposed by Munoz and Mãna in [21] focuses on certifying cloud-based systems using trusted computing platforms. In [13], a security certification framework is proposed. The framework relies on the definition of security properties to be certified. It then uses a monitoring tool to check periodically the validity of the defined properties. In [1], the OPTET project aims to understand the trust relation between the different stakeholders. OPTET offers methodologies tools and models, which provide evidence-based trustworthiness. Additionally, it puts high emphasis on evidence collection during system development, enriched by monitoring and system adaptation to maintain its trustworthiness. In [11], an extension to the Digital Security Certificate [18] is proposed. This certificate contains machine-readable evidence for each claim about the system quality. The certificates are verified by continuously monitoring the certified properties. Differently from the above works, our approach provides a certification process whose trustworthiness is verified by continuously checking the equivalence between a service model and its implementation. Our approach also supports model adaptation to reflect changes that might happen while services are running.

8   Conclusions

In the last few years, the definition of assurance techniques, increasing the confidence of cloud users that their data and applications are treated and behave as expected, has attracted the research community. Many assurance techniques in the context of audit, certification, and compliance domains have been provided, and often build their activities on service modeling. The correctness of these techniques however suffers by hidden changes in the service models, which may invalidate their results if not properly managed. In this paper, we presented a model-based approach to increase the trustworthiness of cloud service certification. Our approach considers service models at different granularity and verifies them against runtime execution traces, to the aim of evaluating their correctness and, in turn, the validity of the corresponding certificates. We also developed and experimentally evaluated a first prototype of our model verification approach.

Acknowledgments. This work was partly supported by the Italian MIUR project SecurityHorizons (c.n. 2010XSEMLC) and by the EU-funded project CUMULUS (contract n. FP7-318580).

References

1. OPTET Consortium: D3.1 initial concepts and abstractions to model trustworthiness (2013). http://www.optet.eu/project/
2. Anisetti, M., Ardagna, C., Damiani, E.: A certification-based trust model for autonomic cloud computing systems. In: Proceedings of the IEEE Conference on Cloud Autonomic Computing (CAC 2014), London, UK, September 2014
3. Anisetti, M., Ardagna, C., Damiani, E.: A test-based incremental security certification scheme for cloud-based systems. In: Proceedings of IEEE SCC 2015, New York, NY, USA, June–July 2015
4. Anisetti, M., Ardagna, C., Damiani, E., Saonara, F.: A test-based security certification scheme for web services. ACM Trans. Web **7**(2), 1–41 (2013)
5. Antunes, N., Vieira, M.: Enhancing penetration testing with attack signatures and interface monitoring for the detection of injection vulnerabilities in web services. In: 2011 IEEE International Conference on Services Computing (SCC), pp. 104–111, July 2011
6. Ardagna, C., Asal, R., Damiani, E., Vu, Q.H.: From security to assurance in the cloud: a survey. ACM Comput. Surv. **48**(1), 1–50 (2015)
7. Ardagna, C., Jhawar, R., Piuri, V.: Dependability certification of services: a model-based approach. Computing **97**(1), 51–78 (2015)
8. Biermann, A., Feldman, J.: On the synthesis of finite-state machines from samples of their behavior. IEEE Trans. Comput. **C-21**(6), 592–597 (1972)
9. Certification infrastrUcture for MUlti-layer cloUd Services. http://www.cumulus-project.eu/
10. Cimato, S., Damiani, E., Zavatarelli, F., Menicocci, R.: Towards the certification of cloud services. In: Proceedings of the IEEE SERVICES 2013, Santa Clara, CA, USA, June–July 2013
11. Di Cerbo, F., Bisson, P., Hartman, A., Keller, S., Meland, P.H., Moffie, M., Mohammadi, N.G., Paulus, S., Short, S.: towards trustworthiness assurance in the cloud. In: Felici, M. (ed.) CSP EU FORUM 2013. CCIS, vol. 182, pp. 3–15. Springer, Heidelberg (2013)
12. Doelitzscher, F., Reich, C., Knahl, M., Passfall, A., Clarke, N.: An agent based business aware incident detection system for cloud environments. J. Cloud Comput. **1**(1), 1–19 (2012)
13. Egea, M., Mahbub, K., Spanoudakis, G., Vieira, M.R.: A certification framework for cloud security properties: the monitoring path. In: Felici, M., Fernández-Gago, C. (eds.) A4Cloud 2014. LNCS, vol. 8937, pp. 63–77. Springer, Heidelberg (2015)

14. Ernst, M., Cockrell, J., Griswold, W., Notkin, D.: Dynamically discovering likely program invariants to support program evolution. In: Proceedings of the 21st International Conference on Software Engineering, ICSE 1999, pp. 213–224. ACM, New York, NY, USA (1999)

15. Fu, X., Bultan, T., Su, J.: Analysis of interacting bpel web services. In: Proceedings of the 13th International Conference on World Wide Web, WWW 2004, pp. 621–630. ACM, New York, NY, USA (2004)

16. Hudic, A., Tauber, M., Lorunser, T., Krotsiani, M., Spanoudakis, G., Mauthe, A., Weippl, E.: A multi-layer and multitenant cloud assurance evaluation methodology. In: 2014 IEEE 6th International Conference on Cloud Computing Technology and Science (CloudCom), pp. 386–393, December 2014

17. Jianqiang, H., Changguo, G., Huaimin, W., Peng, Z.: Quality driven web services selection. In: IEEE International Conference on e-Business Engineering, 2005, ICEBE 2005, pp. 681–688, October 2005

18. Kaluvuri, S., Koshutanski, H., Di Cerbo, F., Mana, A.: Security assurance of services through digital security certificates. In: 2013 IEEE 20th International Conference on Web Services (ICWS), pp. 539–546, June 2013

19. Lorenzoli, D., Mariani, L., Pezzè, M.: Automatic generation of software behavioral models. In: Proceedings of the 30th International Conference on Software Engineering, ICSE 2008, pp. 501–510. ACM, New York, NY, USA (2008)

20. Merten, M., Howar, F., Steffen, B., Pellicione, P., Tivoli, M.: Automated inference of models for black box systems based on interface descriptions. In: Margaria, T., Steffen, B. (eds.) ISoLA 2012, Part I. LNCS, vol. 7609, pp. 79–96. Springer, Heidelberg (2012)

21. Munoz, A., Mãna, A.: Bridging the gap between software certification and trusted computing for securing cloud computing. In: Proceedings of IEEE SERVICES 2013, Santa Clara, CA, USA, June 2013

22. Pearson, S.: Toward accountability in the cloud. IEEE Internet Comput. **15**(4), 64–69 (2011)

23. Rasheed, H.: Data and infrastructure security auditing in cloud computing environments. Int. J. Inf. Manage. **34**(3), 364–368 (2014). ISSN: 0268-4012. http://dx.doi.org/10.1016/j.ijinfomgt.2013.11.002, http://www.sciencedirect.com/science/article/pii/S026840121300145X

24. Ravindran, K.: Model-based engineering methods for certification of cloud-based network systems. In: 2013 Fifth International Conference on Communication Systems and Networks (COMSNETS), pp. 1–2, January 2013

25. Spanoudakis, G., Damiani, E., Mana, A.: Certifying services in cloud: the case for a hybrid, incremental and multi-layer approach. In: 2012 IEEE 14th International Symposium on High-Assurance Systems Engineering (HASE), pp. 175–176, October 2012

26. Sunyaev, A., Schneider, S.: Cloud services certification. Commun. ACM **56**(2), 33–36 (2013)

27. Wu, C., Lee, Y.: Automatic saas test cases generation based on soa in the cloud service. In: 2013 IEEE 5th International Conference on Cloud Computing Technology and Science, pp. 349–354 (2012)

Author Index

Printed in the United States
By Bookmasters